Time Maps

Time Maps

Collective Memory and the Social Shape of the Past

Eviatar Zerubavel

THE UNIVERSITY OF CHICAGO PRESS

CHICAGO AND LONDON

The University of Chicago Press, Chicago 60637
The University of Chicago Press, Ltd., London
© 2003 by The University of Chicago
All rights reserved. Published 2003
Paperback edition 2004
Printed in the United States of America

12 11 10 09 08 07 06 05 04   2 3 4 5
ISBN 0-226-98152-5 (cloth)
ISBN 0-226-98153-3 (paperback)

Library of Congress Cataloging-in-Publication Data

Zerubavel, Eviatar.
    Time maps : collective memory and the social shape of the past /
Eviatar Zerubavel.
        p.   cm.
Includes bibliographical references and index.
    ISBN 0-226-98152-5 (hardcover : alk. paper)
1. Time.   2. History—Philosophy.   3. Civilization—Philosophy.
I. Title.
    BD 638.Z48   2003
    304.2'3—dc21   2002012327

In memory of my grandmother and mother,
my direct links to the past

Contents

Figures

Preface

Most of my work revolves around the study of social structures and patterns, but thus far it has been split along two distinct foci, namely time and cognition. This book is an attempt to integrate those two strands of my scholarly endeavor. On the one hand, it is a continuation of my examination of how we structure time, as presented in my first three books (*Patterns of Time in Hospital Life, Hidden Rhythms,* and *The Seven-Day Circle*) as well as the most recent one (*The Clockwork Muse*). At the same time, it is an extension of work I have done in three other books (*The Fine Line, Terra Cognita,* and *Social Mindscapes*), which analyze thought patterns. By looking at sociomental representations of the past, I thus try to bring these two seemingly disparate facets of my scholarship more closely together.

Studying the past has always fascinated me. Already as a ten-year-old I enjoyed compiling biblical genealogies and lists of ancient monarchs. Books about history, both fictional and nonfictional, have long constituted a major part of my intellectual nourishment. Indeed, many of my grade-school classmates expected me to become a historian when I grew up.

In my earlier work on time I have in fact delved repeatedly into the past, examining critical historical events such as the introduction of the seven-day week, the calendrical separation of Easter from Passover, the invention of the daily schedule, and the introduction of standard time, yet never did I venture to consider history itself as my main object of inquiry. Taking this new step was the result of a highly evocative personal encounter with Mexico in 1984, when, under the irresistible spell of the spectacular ruins of Teotihuacán, Palenque, and Chichén Itzá, I decided to study the social construction of historical continuity and discontinuity. In 1994 I started teaching a graduate seminar, "Time, History, and Memory," and by late 1998 I began working on this book.

My interest in our outlook on the past has been greatly influenced by growing up in Israel, a country deeply obsessed with its history. Yet it was the work of my wife, Yael, on Zionist historiography—culminating in *Recovered Roots,* which I consider the best study of collective memory—that made me realize the tremendous potential of exploring how we collectively envision the past. Her continual encouragement as well as her extremely useful comments on earlier drafts have certainly helped make this a much better book.

My friends and colleagues Paul DiMaggio and Dan Ryan offered me excellent feedback on an early draft of the manuscript. I also benefited from many helpful comments provided by Jim Jasper, John Gillis, Jenna Howard, Karen Cerulo, John Martin, Ruth Simpson, Ann Mische, and Israel Bartal. The great enthusiasm of my editor, Doug Mitchell, with whom my intellectual bond dates to our work together on *Hidden Rhythms* twenty-two years ago, was a tremendous boost during the final stages of completing the book.

As is evident from its title, the book invokes rich topographic and cartographic imagery, which reflects my deep interest in the visual representation of the quasi-spatial features of time. Increasingly fascinated by the prospect of representing my ideas graphically, I soon began to draw on the excellent insights of my son, Noam, who became my special graphic consultant. I am very grateful for the many hours during which he patiently helped me realize my great desire: to depict how we actually map the way time flows in our minds.

The Social Structure of Memory

Why do we think of the Roman Empire as having come to an end in
AD 476 despite the fact that it actually lasted for another 977 years in
Byzantium? Why are racists so obsessed with origins? At what historical
point should the narrative of the conflict between Serbs and Albanians
over Kosovo begin? How did the last shah of Iran manage to spin a 2,500-
year symbolic thread connecting him to Persia's first king, Cyrus, despite
the embarrassing fact that the Pahlavi "dynasty" extended back only one
generation, to his father? Was the tenth century actually less "eventful"
than the twentieth?

By the same token, why did Hernán Cortés practically raze the Aztec
city of Tenochtitlán before proceeding to build Mexico City on its ruins?
Why was the Treaty of Versailles signed in the very same hall where the
German Empire had been formally proclaimed almost fifty years earlier,
following Prussia's 1871 victory over France? Why do six of Angola's seven
national commemorative holidays revolve around its struggle for independence from Portugal during the 1960s and 1970s? Why do Spaniards regard the late-medieval Christian victories over the Moors as a reconquest?
Why do some societies name their children after dead ancestors? Is a thirty-
seventh cousin still a cousin?

To answer such questions we must first examine the unmistakably social maplike structures in which history is typically organized in our minds.
What we need, in other words, is a sociomental topography of the past.

* * *

A "socio*mental* topography" implies a pronouncedly cognitive focus, and this book indeed looks at how the past is registered and organized *in our minds*. I am thus much less concerned with what Jesus, Columbus, or Nebuchadnezzar actually did than with their roles as "figures of memory."[1] In other words, I am primarily interested not in what actually happened in history but in how we *remember* it.

As we very well know, not everything that happens is preserved in our memory, as many past events are actually cast into oblivion. Even what we conventionally consider "history" and thereby include in our history textbooks is not a truly comprehensive record of everything that ever happened, but only a small part of it that we have come to preserve as public memory.

Yet while I definitely do not wish to examine here what actually happened in history, it is also not my intent to simply replace the historian's traditional concern with facts with the psychoanalyst's traditional interest in individuals' idiosyncratic reconstructions of those facts. While the study of memory is quite distinct from the study of what actually happened in the past, it need not be reduced to Rashomonesque personal accounts of individuals. Although memory is not a mere reproduction of objective facts, this does not mean that it is therefore entirely subjective.

Consider the current curricular wars between Eurocentrists and multiculturalists over the literary tradition into which young Americans ought to be socialized or similar cultural battles over women's place in U.S. history. The very existence of such discord reminds us that our recollections of the past are by no means objective, as we clearly do not all remember it the same way. Yet the fact that such *mnemonic battles*[2] usually involve entire groups and are typically fought in unmistakably public forums such as museums and school boards seems to suggest that they are not entirely personal either.

A *socio*mental topography of the past helps highlight this pronouncedly social dimension of human memory by revealing how entire communities, and not just individuals, remember the past. The phenomenology of history it provides is thus grounded in a *sociology* of memory.

In transcending strictly personal recollections, the sociology of memory effectively foregrounds what we come to remember *as social beings*. While there are many memories that we share with no one else, there are specific recollections that are commonly shared by entire groups. One's memories as a Pole, Mormon, or judge, for example, are clearly not just personal.

Unlike psychology, sociology is particularly attentive to the social context within which we access the past, thereby reminding us that we actually remember much of what we do only as members of particular communities. It is thus mainly as a Jew that I remember the destruction of the First Temple more than twenty-five centuries before I was born, and as a track fan that I likewise recall Paavo Nurmi's heroics at the 1924 Olympics.

Being social presupposes the ability to experience things that happened to the groups to which we belong long before we even joined them as if they were part of our own personal past. Such an ability is manifested in the Polynesian use of the first-person pronoun when narrating one's ancestral history[3] as well as in statements like "*I* smelted iron in Nubia" or "*I* built Timbuctoo" used to express a Barbadian poet's distinctly African memories.[4] It is likewise captured in the traditional Jewish belief, repeated every Passover, that "*we* were slaves to Pharaoh in Egypt, and God brought *us* out of there with a mighty hand" and that "in every generation a man should see himself as though *he* had gone forth from Egypt." Such a remarkable existential fusion of one's personal history with that of the communities to which one belongs also helps explain the tradition of pain and suffering carried by American descendants of African slaves as well as the personal sense of shame felt by many young Germans about the atrocities of a regime that ended long before they were born.

Indeed, acquiring a group's memories and thereby identifying with its collective past is part of the process of acquiring any social identity, and familiarizing members with that past is a major part of communities' efforts to assimilate them. Prestigious law firms and elite military units thus usually introduce new members to their collective history as part of their general orientation, and children whose parents came to the United States from Honduras or Laos are nevertheless taught in school to remember the *Mayflower* as part of their new past.[5] By the same token, exiting a social community often involves dispensing with its past; children of assimilated immigrants thus rarely get to learn much from their parents about the history of the societies they chose to physically as well as psychologically leave behind.

Given all this, it comes as no surprise that, when asked to list the names that first come to mind in connection with U.S. history, young Americans often invoke the same historical figures—George Washington, Abraham Lincoln, Thomas Jefferson, Benjamin Franklin.[6] That so many different individuals tend to have the same "free" mnemonic associations suggests that at least some of their seemingly personal recollections may in fact be merely personalized manifestations of a single common *collective memory*.

The memories I examine here are unmistakably collective ones shared by families, ethnic groups, nations, and other *mnemonic communities*.[7] Rather than a mere aggregate of the personal recollections of its various members,[8] a community's collective memory includes only those shared by its members as a group. As such, it invokes a common past that they all seem to recall.

Furthermore, as becomes quite evident on any commemorative holiday, they often recall that past *together*, thereby reminding us that our social environment affects not only what we remember but also *when* we come to remember it! After all, on the same day, an entire mnemonic community manages to focus its attention on the very same moment in history—a remarkable cognitive feat that no other animal has yet been able to accomplish and that makes such holidays truly co-memorative. Such *mnemonic synchronization*[9] was indeed the earliest prototechnological foreglimpse of the modern "global village." On the very same day, the birth of the Prophet is thus jointly remembered by Muslims in Malaysia, Guyana, and Sierra Leone. By the same token, on Good Friday, Christians all over the world come to recall the Crucifixion together, as a single community.

Yet the social nature of human memory is evident not only in the actual content of our recollections but also in the way they are mentally packaged. After all, remembering involves more than just recall of facts, as various mental filters that are quite independent of those facts nevertheless affect the way we process them in our minds (including the way we recall the general gist of past events, which is often all we actually remember of those events),[10] thus leading us to remember some more than others. Such filters are highly impersonal, as they are rarely ever grounded in individuals' own experience. The difference between what Americans and Indians tend to recall from wedding ceremonies,[11] for example, is a product of their having been socialized into different *mnemonic traditions*[12] involving altogether different mental filters commonly shared by their respective mnemonic communities.

Our tendency to better remember facts that fit certain (unmistakably cultural) mental schemata[13] is quite evident in the highly formulaic plot structures[14] we often use for narrating the past. Only in my late thirties, for example, did I first realize that Alfred Dreyfus, whom I had always recalled languishing on Devil's Island (following the infamous trial in which he was wrongly convicted of treason against France) *until he died,* was actually exonerated later by the French authorities and even decorated with the Legion of Honor. Having grown up in Israel and thereby socialized into the Zionist tradition of narrating European Jewish history strictly in terms of persecution and victimhood,[15] such a distorted recollection nevertheless seemed, somehow, to better fit my social schematic expectations.

We normally acquire such habitual mental stances as part of the process of learning to remember in a socially appropriate manner. Far from being a strictly spontaneous act, remembering is also governed by unmistakably social *norms of remembrance*[16] that tell us what we should remember and what we should essentially forget. It is through such *mnemonic socialization*[17] that both born-again Christians and recovering alcoholics, for example, learn to include in their autobiographical accounts some earlier period marked by highly formulaic memories of depravity.[18]

A considerable part of our mnemonic socialization takes place in historical museums and social studies classes, whether as explicit normative prescriptions such as "Remember the Alamo"[19] or as implicitly encoded in virtually any history textbook. Yet much of it also occurs in a somewhat more subtle manner, as when we see George Washington's face on one-dollar bills or notice that almost everything is closed on Christmas Day. Moreover, mnemonic socialization take place in less formal settings such as family gatherings, where it typically involves both actual mentoring (through questions designed to help remind children of things they have experienced)[20] and *co-reminiscing* (as parents and children jointly recount events they have experienced together).[21] It is in such situations that we usually learn the socially appropriate narrative forms for recounting the past as well as the tacit rules of remembrance that help separate the conventionally memorable from that which can—or even ought to—be relegated to oblivion.[22] When a young boy returns from a long day spent with his mother downtown and hears her "official" account to their family of what they did there, he is at the same time receiving a tacit lesson in what is conventionally considered memorable and forgettable.

* * *

Given their unmistakably impersonal nature, social memories are by no means confined, like personal recollections, to our own bodies. It was language that freed human memory from having to be stored exclusively in individuals' brains. Once it became possible for people to share their personal experiences with others through communication, such experiences could be preserved as essentially disembodied impersonal recollections even after they themselves were long gone.

Indeed, language allows memories to actually pass from one person to another even when there is no direct contact between them. As traditional mnemonic go-betweens, old people, for example, often link historically separate generations that would not otherwise have mnemonic access to each other. Such *mnemonic transitivity* enables us to preserve memories in the form of oral traditions that are transmitted from one generation to the next within families, college fraternities, and virtually any other community.

Furthermore, since the invention of writing it is actually possible to bypass any oral contact, however indirect, with any future audience.[23] With patient records, for example, physicians' clinical recollections are readily accessible to any other physician or nurse even when they themselves are not available for consultation.[24] That explains the tremendous significance of documents in business (receipts), law (court decisions), diplomacy (treaties), bureaucracy (minutes), and science (lab reports).[25]

Yet the social preservation of memories does not even require any verbal transmission. Portraits, statues, photographs, and videocassettes, for example, represent various efforts to capture the images and sounds of the past and thereby offer posterity visual as well as auditory access to historical figures and events. Indeed, it is through paintings, compact discs, and television footage that we actually recall the coronation of Napoleon, the voice of Enrico Caruso, or the assassination of John F. Kennedy.

Libraries, bibliographies, folk legends, photo albums, and television archives thus constitute the "sites"[26] of social memory as well as some useful means for studying it. So, for that matter, do history textbooks, calendars, eulogies, guest books, tombstones, war memorials, and various Halls of Fame. Equally evocative in this regard are pageants, commemorative parades, anniversaries, and various public exhibits of archaeological and other historical objects.

There are numerous kinds of data sources on which one can thus draw when conducting research on social memory. The more of them we can incorporate in our studies, the richer those studies are likely to be.[27] Pronouncedly eclectic methodologically, the present book draws on these and many other sites of social memory in a conscious effort to provide as broad a picture of this fascinating phenomenon as possible.

* * *

In trying to uncover the sociomental *topography* of the past,[28] the general thrust of my analysis is also unmistakably structural. While most studies of social memory basically focus on the content of what we collectively remember, my main objective here is to identify the underlying *formal* features of those recollections. Following the fundamental "structuralist" claim that meaning lies in the manner in which semiotic objects are systemically positioned in relation to one another,[29] I believe that the social meaning of past events is essentially a function of the way they are structurally positioned in our minds vis-à-vis other events. I am therefore ultimately interested in examining the *structure* of social memory.

Given its pronouncedly structural focus, the book is thus organized around major formal features of the way we collectively remember the past, as each chapter sheds light on different aspects of its sociomental topography. The main themes of the book are thus unmistakably formal: the perceived "density" of history, the "shape" of historical narratives, the social structure of genealogical "descent," the mental segmentation of essentially continuous historical stretches into discrete "periods," highly structured collective mnemonic distortions of actual historical distances, and so on.

I begin the book by examining the conventional schematic formats that help us mentally string past events into coherent, culturally meaningful historical narratives. In chapter 1 I thus review the major formal patterns along which we normally envision time flowing (linear versus circular, straight versus zigzag, legato versus staccato, unilinear versus multilinear), as quite explicitly evident in the general plots ("progress," "decline," "rise and fall") and subplots ("again and again") of the stories through which we usually come to narrate its passage. I then look at the collectively perceived "density" of the past, as typically manifested in the quasi-topographic layout of the mental relief maps produced by the sharp contrast between what we conventionally recall as "eventful" periods and essentially empty historical "lulls."

In the next two chapters I examine the various mnemonic strategies we normally use to help us create and maintain the illusion of historical continuity. In chapter 2 I look at the different types of bridges we build— physical, calendrical, iconic, discursive—in an effort to "connect" the past and the present, thus shedding some light on the role of anniversaries, revivals, ruins, analogies, and souvenirs in helping coagulate essentially noncontiguous patches of history into a single, seemingly continuous experiential stream. Then, in chapter 3, I offer a close-up of one such form of historical bridging as I explore the genealogical structures of ancestry and descent (dynasties, family trees, pedigrees) that we construct in our minds to help us spin the mental threads we envision as linking past and present members of families as well as underlying our collective visions of nations, "races," and even species.

Yet the effort to establish historical continuity is usually offset by the diametrically opposite sociomental process of constructing historical *discon*tinuity. Whereas the former is geared to produce quasi-contiguity between essentially noncontiguous chunks of history, the latter helps transform continuous historical stretches into series of seemingly distinct segments. At the heart of this process, which I examine in chapter 4, are the "watersheds" we collectively envision separating one supposedly discrete historical "period" from the next. As we shall see, periodizing the past also distorts actual historical distances by essentially compressing those within any given "period" while inflating those across the mental divides separating such conventional segments from one another.

One particularly remarkable manifestation of such social "punctuation" of the past is the mental differentiation of the historical from the merely "prehistorical" through the establishment of what we conventionally come to regard as beginnings. In chapter 5 I look at the social construction of historical beginnings by examining how mnemonic communities (nations, organizations, ethnic groups) envision their collective origins as well as how they try to establish territorial and other political rights by claiming historical priority vis-à-vis other groups. Both of these acts, as we shall see, clearly highlight the common mnemonic effort to enhance one's legitimacy by exaggerating one's antiquity.

* * *

Yet aside from its strictly theoretical implications, so clearly reflected in the way I have organized my discussion, the pronouncedly formal-structural thrust of the book has some very important methodological implications as well, as is quite evident from the unmistakably "formal" manner in which I have collected the data for this study.[30]

As in Euclidean geometry, a strictly formal-structural approach presupposes a conscious obliviousness to scale. My goal, after all, is to develop a general framework that would reveal the fundamental structure of social memory at the macrosocial level of nations as well as the intermediate level of organizations and the microsocial level of families. Only by looking at data from as many "levels" of social units of remembrance as possible would we ever be able to notice the striking formal similarity among the ways in which couples, professions, and religions, for example, normally construct their origins.

In fact, identifying the *generic* features of social memory at each of those levels can help us recognize their manifestations at other levels as well. We might learn quite a lot about how nations present their collective past in history textbooks and national museums, for example, from the way companies or institutes feature theirs in their publicity brochures. A strictly formal-structural approach to memory likewise helps us realize that the way states impose statutes of limitations is fundamentally similar to the way banks establish bankruptcy policies and friends let bygones be bygones!

Furthermore, my pronouncedly generic theoretical concerns call for an explicit commitment to *decontextualize* my findings by pulling them out of the culturally and historically specific environments within which I first happen to identify them,[31] since my ultimate goal is to develop a *transcultural* as well as a *transhistorical* perspective on social memory as a generic phenomenon. Whether a particular national calendar I use in my discussion is Uruguay's or Namibia's is thus by and large secondary to my general interest in the generic features of social commemoration that it helps illustrate. I am likewise less concerned with whether a particular "chain" of monarchs I examine was Egyptian or French, or whether they ruled in the seventeenth century or the third millennium BC, than with the fact that it helps me illustrate some formal features of dynasties *in general*.

The book is thus organized around major formal themes that manifest themselves in a wide variety of substantive contexts. Identifying structural resemblances across such different contexts allows us to appreciate how fundamentally similar are the mnemonic battles between Serbs and Albanians over the "original" settlement of Kosovo and anthropologists and molecular biologists over the dating of the evolutionary split between humans and chimps!

Such pronouncedly generic concerns also call for a conscious effort to draw on a substantively broad base of concrete evidence. Ultimately interested in identifying formal mnemonic patterns that transcend any specific context of remembering, I thus illustrate my arguments with specific examples from a particularly wide range of such contexts. Instead of focally confining myself to one specific case study—a rather common tradition in studies of collective memory[32] that has thus far yielded no serious effort to develop an analytic framework that would be generalizable beyond particular societies at specific historic junctures—I therefore draw my evidence from a wide range of cultural as well as historical contexts. I likewise examine a wide variety of specific domains (science, religion, politics) and sites (calendars, chronicles, pedigrees) of social memory. Needless to say, the wider the range of the contexts on which I draw in my analysis, the broader its generalizability.

Yet though my commitment to *cross-contextual* evidence certainly calls for greater substantive variety, I am not interested here in variation, and my deliberate decision to constantly oscillate between widely different contexts is essentially designed to highlight their *common*, rather than distinctive, mnemonic features. When looking at ethnic dress, historic neighborhoods, and wedding anniversaries, for example, I thus focus primarily on their structural equivalence as forms of "bridging" the past and the present. While this does not necessarily entail a universalistic outlook on memory that basically ignores mnemonic variation,[33] it does entail a commitment to focus on commonality rather than variability. My ultimate goal in this book, therefore, is not to explain mnemonic variation but to identify the common generic underpinnings of the social structure of memory.

The Social Shape of the Past

As one can certainly tell by the fact that we do not recall every single thing that has ever happened to us, memory is clearly not just a simple mental reproduction of the past. Yet it is not an altogether random process either. Much of it, in fact, is patterned in a highly structured manner that both shapes and distorts what we actually come to mentally retain from the past. As we shall see, many of these highly schematic mnemonic *patterns* are unmistakably social.

Plotlines and Narratives

In June 1919, as a triumphant France was preparing to sign the Treaty of Versailles, it made the portentous decision to stage the final act of the historical drama commonly known as *revanche* (revenge) in the very same Hall of Mirrors where the mighty German Empire it had just brought to its knees was formally proclaimed almost fifty years earlier, following Prussia's great victory in the 1870–71 Franco-Prussian War. Not coincidentally, an equally pronounced sense of historical drama led a victorious German army twenty-one years later, in June 1940, to hack down the wall of the French museum housing the railway coach in which the armistice formalizing Germany's defeat in World War I had been signed in November 1918, and tow it back to the forest clearing near the town of Compiègne where that nationally traumatic event had taken place and where Germany was now ready to stage France's humiliating surrender in World War II:

> The *cycle of revenge* could not be more complete. France had chosen as the set-
> ting for the final humbling of Germany in 1919 the Versailles Hall of Mirrors
> where, in the arrogant exaltation of 1871, King Wilhelm of Prussia had pro-
> claimed himself Kaiser; so now Hitler's choice for the scene of his moment of
> supreme triumph was to be that of France's in 1918.[1]

Soon after Hitler finished reading the inscription documenting the historic
humiliation of Germany by France in 1918, everyone entered the famous
railcar and General Wilhelm Keitel began reading the terms of surrender
after explicitly confirming the choice of that particular site as "an act of
reparatory justice."[2]

Only within the context of some larger historical scenario,[3] of course,
could either of these events be viewed in terms of "reparation." And only
within the context of such seemingly never-ending Franco-German revenge scenarios can one appreciate a 1990 joke in which the tongue-in-
cheek answer to the question "Which would be the new capital of the
soon-to-be-reunified Germany: Bonn or Berlin?" was actually "Paris"!

Essentially accepting the structuralist view of meaning as a product of
the manner in which semiotic objects are positioned relative to one another,[4] I believe that the historical meaning of events basically lies in the
way they are situated in our minds vis-à-vis other events. Indeed, it is their
structural position within such *historical scenarios* (as "watersheds," "cat-
alysts," "final straws") that leads us to remember past events as we do.
That is how we come to regard the foundation of the State of Israel, for
example, as a "response" to the Holocaust, and the Gulf War as a belated
"reaction" to the U.S. debacle in Vietnam. It was the official portrayal of
the 2001 military strikes in Afghanistan as "retaliation" for the September
11 attack on the World Trade Center and the Pentagon that likewise led
U.S. television networks to report on them under the on-screen headline
"America Strikes *Back*," and the collective memory of a pre-Muslim, essentially Christian early-medieval Spain that leads Spaniards to regard the
late-medieval Christian victories over the Moors as a "*re*-conquest" *(recon-
quista)*.

Consider also the case of *historical irony*. Only from such a historical
perspective, after all, does the recent standardization of the Portuguese
language in accordance with the way it is currently spoken by 175 million
Brazilians rather than only 10 million Portuguese come to be seen as ironic.
A somewhat similar sense of historical irony underlies the decision made

by the *New York Times* the day after the 2001 U.S. presidential inauguration to print side by side two strikingly similar yet contrasting photographs featuring the outgoing president Bill Clinton outside the White House: one with his immediate predecessor, George Bush, back in January 1993, and the other with his immediate successor, George W. Bush, exactly eight years later.[5]

One of the most remarkable features of human memory is our ability to mentally transform essentially unstructured series of events into seemingly coherent *historical narratives*. We normally view past events as episodes in a story (as is evident from the fact that the French and Spanish languages have a single word for both *story* and *history*, the apparent difference between the two is highly overstated), and it is basically such "stories" that make these events historically meaningful. Thus, when writing our résumés, for example, we often try to present our earlier experiences and accomplishments as somehow prefiguring what we are currently doing.[6] Similar tactics help attorneys to strategically manipulate the biographies of the people they prosecute or defend.

As is quite evident from figure 1, in order for historical events to form storylike narratives, we need to be able to envision some connection between them. Establishing such unmistakably contrived connectedness is the very essence of the inevitably retrospective mental process of *emplotment*.[7] Indeed, it is through such emplotment (as well as reemplotment,[8] as is quite spectacularly apparent in psychotherapy)[9] that we usually manage to provide both past and present events with historical meaning.

Approaching the phenomenon of memory from a strictly formal narratological perspective, we can actually examine the structure of our collective narration of the past just as we examine the structure of any fictional story.[10] And indeed, adopting such a pronouncedly morphological stance helps reveal the highly schematic formats along which historical narratives usually proceed. And although actual reality may never "unfold" in such a neat formulaic manner, those scriptlike *plotlines* are nevertheless the form in which we often remember it, as we habitually reduce highly complex event sequences to inevitably simplistic, one-dimensional visions of the past.

Following in the highly inspiring footsteps of Hayden White,[11] I examine here some of the major plotlines that help us "string" past events in our minds,[12] thereby providing them with historical meaning. Rejecting, however, the notion that these plotlines are objective representations of

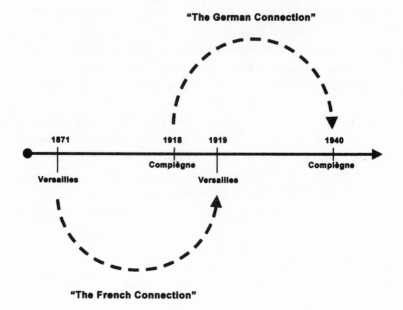

Figure 1 *The Versailles and Compiègne Plotlines*

actual event sequences, as well as the assumption that such visions of the past are somehow universal, I believe that we are actually dealing here with essentially conventional *socio*mnemonic structures. As is quite evident from the fact that certain schematic formats of narrating the past are far more prevalent in some cultural and historical contexts than others, they are by and large manifestations of unmistakably social traditions of remembering.

Progress

A perfect example of such a plotline is the general type of historical narrative associated with the idea of *progress*. Such a "later is better" scenario is quite commonly manifested in highly schematic "rags-to-riches" biographical narratives[13] as well as in unmistakably formulaic recollections of families' "humble origins." It can likewise be seen in companies' "progress reports" to their shareholders as well as in history of science narratives, which almost invariably play up the theme of *development*.

Yet the most common manifestation of this progressionist[14] historical scenario is the highly schematic backward-to-advanced evolutionist narrative. It is quite evident, for example, in conventional narrations of human origins, which typically emphasize the theme of progressive *improvement* with regard to the "development" of our brain, level of social organization, and degree of technological control over our environment. Similarly, it is evident whenever modern, "civilized" societies are compared to so-called underdeveloped, "primitive" ones.[15]

As we can see in figure 2, such an unmistakably schematic vision of progressive improvement over time often evokes the image of an upward-leaning ladder. This common association of time's arrow with an upward direction (and its rather pronounced positive cultural connotations)[16] is quite crisply encapsulated in the title of Jacob Bronowski's popular book and television series, *The Ascent of Man*,[17] as well as in the conventional vision of the "lower" forms of life occupying the lower rungs of the "evolutionary ladder."[18]

Such a highly formulaic vision of the past clearly reflects more than just the way some particularly optimistic individuals happen to recall certain specific events. Indeed, it is part of the general historical outlook of entire mnemonic communities. Though we normally regard optimism as a personal trait, it is actually also part of an unmistakably schematic "style" of remembering shared by entire communities.

Thus, as is quite evident from Horatio Alger's and numerous other "rags-to-riches" versions of the so-called American Dream, many Americans, for instance, are much greater believers in the idea of progress than Afghans or Australian Aborigines. And as one can clearly tell from the general aversion of the working class to this idea,[19] different historical outlooks are also associated with different social classes.[20]

Furthermore, as a brainchild of the Enlightenment, progressionism is a hallmark of modernity and has certainly been a much more common historical outlook over the past two hundred years than during any earlier period. Viewing history in terms of progress is an integral part of the late eighteenth- and early nineteenth-century philosophies of Marie Jean Condorcet, Georg Wilhelm Friedrich Hegel, and Auguste Comte. It is likewise encapsulated in major late nineteenth-century offshoots of those philosophies such as the social and cultural evolutionism of Herbert Spencer, Lewis Henry Morgan, and Edward B. Tylor, who basically envisioned human history as a progressive ascent from savagery to civilization.[21]

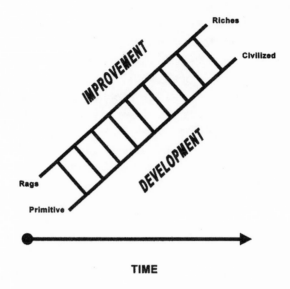

Figure 2 *The Progress Narrative*

Decline

This essentially forward-looking view of history sharply contrasts with yet another conventional historical outlook, which basically features *decline* as the major theme in accordance with which we come to organize our memory.[22] Inherently pessimistic, this unmistakably backward-clinging historical stance typically includes an inevitably tragic vision of some glorious past that, unfortunately, is lost forever. In marked contrast with the progress narrative, in the decline narrative things usually get worse with time. Instead of improvement, this essentially regressive[23] mnemonic tradition emphasizes *deterioration,* thereby promoting a general view of the past most effectively represented by a downward-pointing arrow, as in figure 3. No wonder it is often coupled with a deep sentimental attachment to "the good old days."[24] Whereas progress implies an idealized future, *nostalgia* presupposes a highly romanticized past.

Note, however, that we are not dealing here with actual historical trends but with purely mental historical outlooks. The very same historical period, after all, is remembered quite differently, depending on whether we use a progress or a decline narrative to recount it. During the 1992 U.S. presiden-

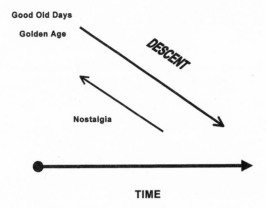

Figure 3 *The Decline Narrative*

tial election, for example, while George Bush was portraying his presidency as a period of substantial progress marked by the downfall of communism and the emergence of a new U.S.-dominated world order, a very different picture was being presented by his challenger, Bill Clinton, who quite effectively downplayed those historic international developments by relentlessly focusing on the alarming rise in domestic poverty and unemployment.

As exemplified by parole hearings and tenure reviews, historical plotlines are often extrapolated to imply *anticipated* trajectories. To appreciate such inherently strategic manipulation of decline narratives, consider a provocative display of the devastating effects of deforestation at Costa Rica's Lankester botanical gardens in Cartago. A series of maps depicting the progressively decreasing amount of Costa Rican land still covered by rain forest are sequentially arranged to form a disturbing narrative that begins in 1940 with an almost entirely green country and ends in the year 2025, quite evocatively represented by a virtually empty map with a big question mark. As one might expect, projecting such historical regression onto the future is a major feature of "doomsday" scenarios.

Often articulated in nostalgic visions of some mythical golden age after which things have essentially been going "downhill," such a pronouncedly regressive mnemonic tradition is also quite apparent in the general tendency to remember our ancestors as larger-than-life, almost superhuman figures. Such an inherently conservative historical outlook, succinctly encapsulated in the traditional Jewish belief that every generation is of a somewhat lesser quality than its predecessors *(holekh u-fochet ha-dor)*,[25] is explicitly manifested in the divine pedigree ascribed to humankind in various cosmogonies. It is also implicit in the unmistakably downward direction in which we conventionally depict the flow of time in family trees and other maps of so-called descent[26] as well as in the highly reverential manner in which we normally think about Shakespeare or Mozart, or the way we tend to remember our national "Founding Fathers" as well as past sports "legends."

Like its progressionist counterpart, this highly formulaic vision of the past represents a particular social tradition of remembering. Though we normally regard pessimism, like optimism, as a personal trait, actually it is also part of an unmistakably schematic style of remembering shared by entire mnemonic communities. Indeed, though "virtually every culture past or present has believed that men and women are not up to the standards of their parents and forebears,"[27] this particular view of the past (just like nostalgia)[28] is much more common in some historical periods than in others. And although the vision of our tragically irretrievable Edenic origins dates to ancient Judaism and our progressive *degeneration* from some idealized golden age was already recounted by Hesiod 2,700 years ago, many decline narratives are in fact a reaction to the overly optimistic modern belief in progress. This is quite evident in the highly pessimistic philosophies of Arthur Schopenhauer and Friedrich Nietzsche as well as in the unmistakably modern social and biological degeneration narratives produced by Cesare Lombroso, Edwin Ray Lankester, Max Nordau, and Oswald Spengler.[29]

A Zigzag in Time

Despite the obvious difference between them, however, both progress and decline narratives share one important formal feature. Whether their basic underlying plotline points upward or downward, the overall story it entails has a single, unmistakably uniform direction. The situation is quite different in narratives that specifically combine upward- and downward-pointing plotlines in an effort to highlight significant *changes* in historical trajectories. Instead of featuring just progress or decline, these narratives feature both.

As one might expect, such *"zigzag" narratives* assume one (or some combination) of two basic forms. One is the *rise-and-fall narrative*, an essentially tragic scenario in which, following some unfortunate event such as losing one's job, going bankrupt, or losing a war, a story of success suddenly turns into one of decline. The histories of the Roman, British, and Ottoman Empires or the high-tech industry in the 1990s are some classic examples of this highly formulaic narrative. The other, essentially obverse form is the Cinderella-like *fall-and-rise narrative,* in which a sharp descent is suddenly reversed, thereby changing to a major ascent. A perfect example is the *conversion narrative,* in which a moral decline is finally brought to a happy end through the discovery of some new source of spiritual light, as in the case of "born-again" Christians,[30] or the *recovery narrative* so common in clinical rehabilitation programs such as Alcoholics Anonymous, where members are expected to actually "hit bottom" before they can begin their ascent back to well-being.[31] Such a highly schematic pattern is typical of autobiographical narratives that involve dramatic rebounds following some major decision to quit smoking, get a divorce, or go back to school. It is also evident in narratives of national redemption, such as the conventional postwar economic recovery histories of Germany and Japan.

Both rise-and-fall and fall-and-rise narratives, however, share an important formal feature, which is that they always involve some dramatic *change of course.* Whether the critical turn is upward or downward, it essentially entails a major redirection of a historical trajectory,[32] sometimes even a complete reversal. *Turning points* are the mental road signs marking such perceived transitions.[33]

Given that such "changes of course" are, after all, only mental constructions, we should not be surprised to find significant differences in how various mnemonic communities come to remember any historical "transition." To appreciate such sociomnemonic pluralism, compare the highly divergent Eastern European memories of the Communist period, or the Democratic and Republican visions of the late 1970s and early 1980s in the United States. For example, in the 1984 presidential election, former vice president Walter Mondale kept presenting his years in office during the late 1970s as a period of considerable social progress that ended with Ronald Reagan's 1980 election victory, which basically led the United States into a downward path that could only be reversed if he, Mondale, were elected president. As we can see in figure 4, he was thus invoking a classic rise-and-fall scenario with a possible future "happy ending" coda by associating 1980 with the onset of a four-year period of sharp decline and identifying 1984 as the potential beginning of a new period of resumed

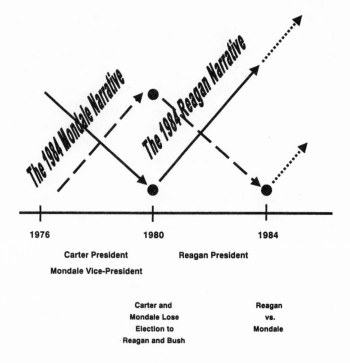

Figure 4 *Democratic and Republican Visions of 1976–84*

progress. Reagan, on the other hand, was invoking a diametrically opposite fall-and-rise scenario featuring 1980 as a critical turning point upward, essentially reversing the disastrous political and economic downslide of the Carter-Mondale years. Confidently prompting voters to compare their quality of life in 1980 with their situation in 1984, he thus presented his first term in office as a period of great progress that would continue for at least "Four More Years" if he were reelected.

Ladders and Trees

Whether they are about progress, decline, or some "zigzag" combination of both, all the historical narratives we have thus far examined essentially involve strictly unilinear plotlines. Yet those are by no means the only kinds of trajectories in which we normally organize the past in our minds.

As aptly captured in the aforementioned ladder metaphor,[34] the essence of unilinearity is the vision of a serial progression, a one-dimensional sequence of unmistakably *successive* episodes such as the Stone Age, Bronze Age, and Iron Age; the 1950s, 1960s, and 1970s; or childhood, adulthood, and old age. This vision is explicitly articulated in the so-called phyletic model of organic evolution as well as in the very idea of the "life *course*." It is also endemic to cultural evolutionism, which basically places all human cultures, past and present, on different rungs of the same ladder. Such a perspective inevitably implies rejecting the very possibility of any form of cultural contemporaneity, to the point of actually regarding the "primitive" as some past version of the modern. "Lower" cultures are thus seen as living fossils, essentially frozen relics of *our* ancient past![35]

Inherently teleological, *unilinear narratives* often attribute some purposeful design to history.[36] As such, they usually also regard the overall direction of the historical trajectories they describe as largely predetermined. Like an escalator, history is thus seen in such narratives as having a clear course that is often articulated in terms of general laws.[37] According to Auguste Comte, for instance, these highly *deterministic* laws actually dictate the path through which the human mind basically *has to* proceed as it moves from one stage of historical development to the next, since each of those stages is "the necessary result of the preceding, and the indispensable mover of the following."[38]

An early proponent of cultural evolutionism, Comte repeatedly invoked "evolution,"[39] a concept most probably borrowed from embryology, a particular branch of biology explicitly centered on a seemingly predetermined process of evolving.[40] Indeed, *evolutionary narratives* are essentially teleological stories of "becoming." This is quite evident in cultural evolutionist narratives, which tend to portray modern civilization as the epitome of social, political, and economic "development,"[41] as well as in biological ones, which basically consider humans the pinnacle of creation and the entire three-billion-year evolution of life on this planet a single monothematic story leading to its "final product." Essentially regarding "lower" forms of life as mere stages in the pronouncedly unilinear evolution of "higher" ones, such narratives thus view apes, for example, as the products of early failed attempts to create man![42]

Ironically, the reality of such "failed experiments" has actually led us to back away from unilinearity and develop an altogether different form of narrating the past. It was his growing awareness of biological extinction that led Georges Cuvier two centuries ago to introduce *multilinear* historical narratives[43]—and their awareness of hominid extinctions in particular that convinced anthropologists later to follow his lead. The realization that some Neanderthals actually lived alongside (rather than only before) anatomically modern humans and thus could not have been our direct ancestors first made us aware of the inevitable historiographical implications of such extinctions.[44] Viewing the Neanderthals as a "dead-end" branch of our family tree[45] led to our abandonment of the unilinear vision of human evolution—which, given its inevitably simplistic, one-dimensional image of successive species essentially *replacing* one another, obviously cannot account for such seemingly "anachronistic" contemporaneity. Only the stubborn refusal to accept the fact that some hominid species (and not just dodoes and dinosaurs) actually died out without issue still prevents some of us from accepting the idea of multilinearity.[46]

Such essentially anthropocentric blinders also prevent some of us from fully appreciating the nonteleological, unmistakably *contingent* nature of organic evolution (or any other historical process, for that matter) that is inevitably implied in a multilinear narration of history. The fact that in the overall drama of evolution our "star actor" has actually been "offstage for 99.99 percent of the play"[47] should help us recognize that evolution is an essentially purposeless, haphazard process that does not necessarily lead to humankind. Indeed, many of the ancient fossils we find today actually lie entirely off the direct ancestral path to us.[48]

It was probably August Schleicher's ingenious use of *cladograms* (branching diagrams) in the 1850s to represent the complex genealogical relations between different languages[49] that inspired Charles Darwin to present his essentially multilinear narrative of the evolution of life in the form of a tree, with the bifurcation of species along ever-diverging branches (speciation) playing a critical role in that process.[50] Following Darwin, most biologists today seem to prefer the image of a two-dimensional tree to that of a one-dimensional ladder for representing this remarkably complex process. As a result, we now envision life as "a copiously branching bush, continually pruned by the grim reaper of extinction, not a ladder of predictable progress."[51]

Such pronouncedly multilinear imagery also helps remind us that " 'simpler' creatures are not human ancestors . . . but only collateral branches on life's tree,"[52] since, after all, "no living species can be the ancestor of any other."[53] As is evident from the cladogram in figure 5, despite the various popular graphic representations of human evolution inspired by unilinear narratives,[54] modern chimps and gorillas are not

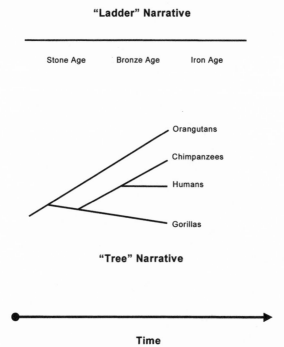

"Ladder" Narrative

Stone Age   Bronze Age   Iron Age

Orangutans

Chimpanzees

Humans

Gorillas

"Tree" Narrative

Time

Figure 5 *Unilinear and Multilinear Historical Plotlines*

"early" forms of human evolution but our own *contemporaries*![55] Cultural evolutionism notwithstanding, that is also true of "primitive" cultures.[56]

Circles and Rhymes

In every historical narrative we have thus far examined, unilinear and multilinear alike, time always seems to be moving "forward." Within any sequence of events we remember, therefore, it is always quite clear which ones occurred earlier and which ones only later. However, there is yet one other major schematic form of organizing our memories that presupposes no such directionality.

Although we usually view time as an entity that can be graphically represented by a straight arrow, as in figures 1 through 5, we sometimes also experience things as moving "in circles."[57] (These two contrasting visions

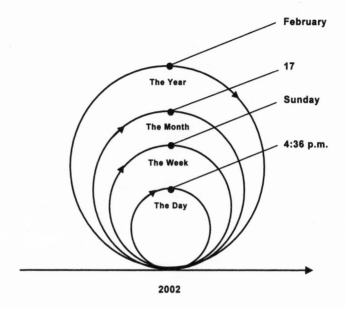

Figure 6 *Linear and Circular Visions of Time*

of time are not incompatible, however. As we can see in figure 6, locating a particular historical instant in 2002, for example, does not preclude it from also being designated as 4:36 P.M. on Sunday, 17 February, thereby placing it on four different wheels that are nevertheless rolling along an unmistakably straight road.)[58] Essentially rejecting the linear vision of historical events as unique occurrences, such a distinctly *cyclical* view of history basically envisions things as being trapped, like the main protagonist of the movie *Groundhog Day*, in some eternal present. Jews' traditional identification of their enemies as Amalek is thus not just metaphorical: within such mythical "panchronistic"[59] vision, that wretched biblical enemy is virtually still alive! After all, our distinctly modern notion of *anachronism* does not even exist within such a pronouncedly nonlinear view of history.[60]

As odd as it may seem to us now, until relatively recently that was the way humans had probably always experienced time. Only in the last couple of millennia, in fact, did our uncompromisingly linear view of the past—symbolically captured in the modern relegation of "time travel" to science fiction—actually come into being. Yet though we may have virtually abandoned the mythical belief that history actually repeats itself,[61] we have nevertheless preserved a somewhat milder version of this traditional nonlinear vision of time.

The essence of this version is quite poignantly captured by the quip, attributed to Mark Twain, that "history doesn't repeat itself, but it rhymes."[62] As we can see in figure 7, such *historical "rhyming"* is what actually enables us to envision cycles. While reading each of the three "poems" horizontally offers us a strictly linear view of history, reading them vertically allows us to notice recurrence (of autumns, Saturdays, and presidential elections). Such "rhyming" implies that, while clearly distinct, the past and the present are nonetheless fundamentally similar, to the point of evoking a déjà-vu sense of "there we go *again*."

As exemplified by such *recurrence narratives,* memory often schematizes history by essentially "fusing analogous personalities or situations into one."[63] The Rwandan mnemonic tradition of clustering past monarchs in cycles of four in accordance with the unmistakably formulaic pattern "A conquers, B is unlucky, C prospers, and D is a legislator"[64] is a perfect case in point. That people actually recall a particular king as a "conqueror" or "legislator" reminds us that social memory basically consists of not only specific historical figures (Innocent III) and events (the Crimean War) but also distinctly generic types of figures (popes) and events (wars).[65]

Such *mnemonic typification* is particularly evident when we mistake one specific historical figure or event for another. Inherently intracategorical, these *mnemonic slips* help reveal the outlines of the conventional categories in which we tend to mentally lump "similar" historical figures or events together.[66] The typified manner in which Israelis, for example, come to remember their national past becomes quite apparent when they confuse traditional holidays designed to commemorate distinct historical events yet nevertheless involving the very same schematic formula ("military uprising against foreign occupation").[67] Mnemonic typification is also remarkably apparent when we happen to recall in great detail something that happened to one of our children, yet fail to remember exactly which one!

Mountains and Valleys

Aside from their overall trajectories, however, historical narratives also vary considerably in their perceived "density." Equally critical in affecting the

Winter 1973, Spring 1973, Summer 1973, **Autumn** 1973
Winter 1974, Spring 1974, Summer 1974, **Autumn** 1974
Winter 1975, Spring 1975, Summer 1975, **Autumn** 1975
Winter 1976, Spring 1976, Summer 1976, **Autumn** 1976

Sunday June 2,	Monday June 3,	Tuesday June 4,	Wednesday June 5,	Thursday June 6,	Friday June 7,	**Saturday** June 8
Sunday June 9,	Monday June 10,	Tuesday June 11,	Wednesday June 12,	Thursday June 13,	Friday June 14,	**Saturday** June 15
Sunday June 16,	Monday June 17,	Tuesday June 18,	Wednesday June 19,	Thursday June 20,	Friday June 21,	**Saturday** June 22
Sunday June 23,	Monday June 24,	Tuesday June 25,	Wednesday June 26,	Thursday June 27,	Friday June 28,	**Saturday** June 29

1985 (Post-election year), 1986 (Midterm Elections), 1987 (Pre-election year), 1988 **(Election year)**
1989 (Post-election year), 1990 (Midterm Elections), 1991 (Pre-election year), 1992 **(Election year)**
1993 (Post-election year), 1994 (Midterm Elections), 1995 (Pre-election year), 1996 **(Election year)**
1997 (Post-election year), 1998 (Midterm Elections), 1999 (Pre-election year), 2000 **(Election year)**

Figure 7 Historical Rhymes

general shape of these narratives, such *mnemonic density* reflects how intensely we actually remember different historical periods.

As a strictly mathematical entity,[68] time is homogeneous, with every minute essentially identical to every other minute, as demonstrated by the way they are conventionally measured by the clock. Experientially, however, minutes vary considerably depending on whether we are aroused or bored, whether our favorite team is leading or losing, and so on.[69] Yet the different qualities we attach to time are not just personal. As exemplified by the much higher rate at which we are paid for the same amount of work time if it is officially considered "overtime,"[70] equal durations are often made unequal *socially*.[71] Just as we conventionally distinguish "holy" days from the seemingly characterless intervals between them,[72] such qualitative heterogeneity[73] is epitomized by the way we differentiate extraordinary *("marked")* from mere ordinary *("unmarked")* time[74]—perfectly exemplified by the week, a cycle of periodically alternating "marked" and "unmarked" days specifically designed to signify major cultural contrasts between ordinary and extraordinary chunks of social reality.[75]

This pronouncedly *qualitative approach to time* is also evident in the way we envision the past, the social shape of which is profoundly affected by the rather pervasive sociomental differentiation of *"eventful"* historical periods from *"uneventful,"* seemingly empty historical "lulls."[76] Generally regarded as less memorable, "unmarked" stretches of history are essentially relegated to social oblivion. As a result, we come to remember some historical periods much more intensely than others. A powerful social projector thus highlights certain parts of the past while basically leaving others in total darkness[77]—which is precisely how we have come to regard the seemingly uneventful centuries between the collapse of the Mycenaean civilization circa 1100 BC and the rise of the classical Hellenic world circa 800 BC as a "dark" age.[78]

Such an inherently "optical" vision of the past is a product of certain *norms of historical focusing* that dictate what we should mnemonically "attend"[79] and what we can largely ignore and thereby forget. It thus basically involves a fundamental distinction (closely resembling the one between "figure" and "ground")[80] between what we regard as historically "significant" and thus come to collectively remember, and what is considered "irrelevant" and thereby essentially relegated to social oblivion.[81] The common tendency to regard wars as eventful and thus memorable, yet the considerably longer "quiet" periods between them as practically empty, is a perfect case in point.

As demonstrated by the fact that we can actually envision even several consecutive centuries as virtually empty,[82] historical periods clearly vary in their perceived density. History thus takes the form of a relief map, on the mnemonic hills and dales of which memorable and forgettable events from the past are respectively featured. Its general shape is thus formed by a handful of historically "eventful" mountains interspersed among wide, seemingly empty valleys in which nothing of any historical significance seems to have happened.[83]

As this explicitly topographic imagery seems to imply, socially "marked" historical periods clearly occupy much more mnemonic "space" than one would expect on strictly mathematical grounds. This variable density of historical intervals constitutes a significant semiotic code. As Claude Lévi-Strauss has noted,

> We use a large number of dates to code some periods of history, and fewer for others. This variable quantity of dates applied to periods of equal duration are a gauge of what might be called *the pressure of history*: there are . . . periods where . . . numerous events appear as differential elements; others, on the contrary, where . . . (although not of course for the men who lived through them) very little or nothing took place. . . . Historical knowledge thus proceeds in the same way as a wireless with frequency modulation: like a nerve, it codes a continuous quantity . . . by frequencies of impulses proportional to its variations.[84]

Thus, for example, in the Book of Chronicles the reign of King Solomon is allotted 201 verses, while that of Joash only receives 27 despite the fact that both are reported there to have lasted forty years. King Hezekiah's twenty-nine-year reign is likewise marked much more prominently (117 verses) than the fifty-five-year reign of his son, Manasseh (only 20 verses).[85] That clearly tells us something about the relative place of both Solomon and Hezekiah in Jewish collective memory.

Consider also the relative amount of space allotted in Clifton Daniel's *Chronicle of America* to each of the nearly fifty decades of American history from 1492 to 1988. As we can see in figure 8, the differential mnemonic marking of mathematically identical historical intervals is quite revealing. Especially in contrast with the amount of space allotted to their immediate chronological neighbors (the 1850s and 1950s), the actual number of pages allotted to the 1860s and 1940s, for instance, is quite suggestive of the particular memorability of wartime periods, since from a strictly mathematical standpoint those decades were absolutely identical. By the same token, when the exact same amount of space in the book (twenty-four pages) is allotted to the *three*-year interval from 1775 through 1777 as well as to the *sixty*-year interval from 1690 through 1749,[86] it is quite clear how different those two periods are in terms of their perceived historical "eventfulness" (and therefore social memorability). The fact that most Americans seem to know much more about the 1770s than about the 1830s obviously suggests more than just a matter of recency.

Such essentially qualitative heterogeneity of mathematically identical time intervals underscores a pronouncedly *nonmetrical* approach to chronology that basically involves mnemonically inflating certain historical periods while compressing others. On the unmistakably nonmetrical time lines implicitly encapsulated in elderly Germans' life histories, for example, the years 1935–41 thus seem virtually empty compared to the years 1942–45.[87]

Yet collective memory is more than just an aggregate of individuals' personal memories, and such inevitably personal relief maps cannot possibly capture what an entire nation, for example, *collectively* considers historically eventful or uneventful. To observe the social "marking" of the past, we therefore need to examine *social time lines* constructed by entire mnemonic communities. For that we must turn to unmistakably social sites of memory. As one might expect, historical periods that are allotted more pages in official history textbooks or assigned special wings in national museums are indeed those *sacred periods*[88] on which nations are most intensely focused mnemonically. And since the sacred is often man-

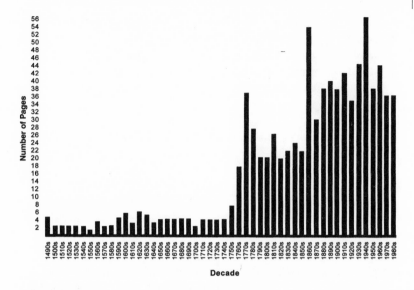

Figure 8 *The Sociomnemonic Density of American History*

ifested in ritual display,[89] we also need to examine the way major figures and events from the past are ritually commemorated. After all, by carving marked periods out of essentially unmarked stretches of history, *ritual commemoration* helps mnemonic communities explicitly articulate what they consider historically eventful. As it "lifts from an ordinary historical sequence those extraordinary events which embody our deepest and most fundamental values," it thus basically serves as "a register of sacred history."[90] Indeed, commemorative rituals[91] often embody major social time lines.

Consider, for example, explicitly commemorative ritual displays such as the postage stamps, street names,[92] and public parades specifically designed to commemorate major historical figures or events. Examining the way they are chronologically distributed certainly helps identify sacred periods in a group's history. As W. Lloyd Warner demonstrated in a classic study of the actual historical contents of a commemorative procession featuring the first three centuries of a New England town's collective past, the events constituting what mnemonic communities come to regard as their history are *unevenly distributed chronologically:*

The forty-three floats of the Procession . . . were spread throughout the three hundred years being officially celebrated. . . . . [C]hronologically they are not spread equally throughout the three centuries. There are sharp divergencies between the social time of the Procession and the chronology of objective time. . . . Since three hundred years were being celebrated, if only the statistical probability of pure chance were at work each century would receive a third of the scenes displayed and each half- and quarter-century be given its proportion of symbolic events. . . . But in fact, one brief period of little more than a decade received as much attention as the previous hundred years. One full quarter-century was not represented at all.[93]

Essentially contrasting metrical "chronology" with unmistakably nonmetrical "social time," he thus proceeded to identify uneven chronological distribution patterns such as having the years 1780–1805 represented by ten floats, yet the mathematically identical period from 1705 to 1730 by virtually none![94] Such patterns have also been observed in a similar examination of the *chronological density* of the historical events that are publicly commemorated in the U.S. Capitol's art collection in Washington, D.C. One only needs to compare the United States' public commemoration of the highly "eventful" 1770s and virtually barren 1760s,[95] for example, to become fully aware of the fundamental contrast between the sacred mountains and profane valleys of the past.

Another extremely useful social site of memory in this regard is the calendar. As a cycle of "holy days" specifically designed to commemorate particular historical events, the calendar year usually embodies major narratives collectively woven by mnemonic communities from their past. Examining which particular events are commemorated on holidays can thus help us identify sacred periods in their history.

For instance, the remarkably "dense" cluster of historical events commemorated every year by Libya on Revolution Day (the overthrow of King Idris by Colonel Muammar Qaddafi), British Bases Evacuation Day (the closing of the military bases at al-Adem and Tobruk), American Bases Evacuation Day (the closing of the Wheelus Air Force Base), and Evacuation of Fascist Settlers Day (the expulsion of Italians from Libya) all occurred during the brief yet exceptionally "eventful" period between September 1969 and October 1970. In like manner, Angola sets aside five days every year—Armed Forces Day, Heroes' Day, Independence Day, Victory Day, and MPLA Foundation Day—to commemorate the three-year period from the resumption of its national struggle for independence from Portugal in 1974 to the eventual transformation of the Popular Movement for the Liberation of Angola into a full-fledged political party in 1977. Note also the extremely disproportionate mnemonic preoccupation with the periods from 1803 to 1805 in Haiti, 1990 to 1991 in Azerbaijan, 1825 to 1828 in Uruguay, 1919 to 1923 in Turkey, and 1896 to 1898 in the Philippines, each of which is specifically commemorated every year on at least three different national holidays.[96]

Essentially housing annual cycles of commemorative holidays, calendars normally entail seismogram-like narratives encapsulating groups' histories in the form of some highly memorable sacred peaks sporadically protruding from wide, commemoratively barren valleys of virtually unmarked, profane time. By highlighting the pronouncedly variable mnemonic density of different stretches of history, these *commemograms* thus capture the uneven chronological distribution of historical "eventfulness."

An extensive cross-national examination of 191 such commemograms[97] reveals a most intriguing pattern. As far as national memory is concerned (although evidence seems to suggest that this is a much more general pattern),[98] the social shape of the past is essentially bimodal, with most of the events commemorated on national holidays having occurred either in the very distant past or within the last two hundred years. Events that are calendrically commemorated by nations thus typically form two chronologically dense clusters representing their respective spiritual and political origins and separated from each other by long stretches of commemoratively "empty" time.

The official sociomnemonic tour of the past formally encapsulated in the national calendar of Thailand perfectly illustrates this rather pervasive pattern. It opens by featuring three major events in the life of the Buddha—his birth circa 563 BC (commemorated annually on Visakha Buja), his first public sermon circa 528 BC (Asalaha Buja), and the announcement of his imminent death circa 483 BC (Makha Buja).[99] As we can see in figure 9, this mnemonically dense eighty-year period is followed by a commemoratively barren 2,265-year historical "lull," which ends in 1782 with the foundation of the current royal dynasty (Chakkri Day). Thailand's three remaining historical holidays are specifically designed to commemorate the reign of King Rama V from 1868 to 1910 (Chulalongkorn Day), the country's historic transition to constitutional monarchy in 1932 (Constitution Day), and the accession of its present ruler, King Rama IX, in 1946 (Coronation Day).[100]

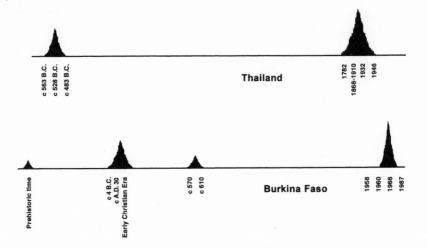

Figure 9 *National Commemograms*

Yet societies often encompass more than just a single mnemonic community, and some countries consequently observe holidays of two (Syria), three (Suriname), four (Bangladesh), and even six (India) different religions, thereby officially commemorating side by side *multiple pasts* that are quite independent of one another. As one might expect, when nations trace their spiritual roots to more than one religion, their calendars often embody commemograms reflecting the structural complexity of their identities.

A fine example of such *mnemonic syncretism* is the "three-act" commemogram encapsulated in Burkina Faso's national calendar. As we can see in figure 9, with the single exception of the thwarted sacrifice of Ishmael, a mythical prehistoric event commemorated annually on the Muslim holiday Tabaski, the former Upper Volta's calendrically commemorated past basically consists of three sacred historical mountains separated from one another by wide, virtually empty historical valleys. A first cluster of commemorative holidays specifically designed to invoke the country's Christian roots features the birth of Jesus circa 4 BC (Christmas), his ascension to heaven circa AD 30 (Ascension Day), his mother's assumption into heaven not too long after that (Feast of the Assumption of the Blessed Virgin Mary), and a chronologically vague early period represented by All Saints' Day. A second cluster designed to invoke Burkina Faso's distinctive Muslim roots features the birth of Mohammed circa AD 570 (Birthday of the Prophet) as well as the period circa 610 when he started having divine revelations (Ramadan). A long, commemoratively empty 1,350-year lull is then followed by a third cluster of relatively recent national political events such as Upper Volta's decision to become a republic in 1958 (Republic Day), its formal independence from France in 1960 (Independence Day), and the military overthrows of presidents Maurice Yaméogo in 1966 (Revolution Day) and Thomas Sankara in 1987 (Anniversary of the 1987 Coup).

One of the most striking features of such commemograms is the long historical stretches that are left virtually empty in groups' collective memories. Thus, throughout the Muslim world, a thirteen-century calendro-commemorative gap extends from Mohammed's celebrated night journey to heaven circa AD 620 (which is traditionally commemorated on Leilat al-Meiraj) or the martyrdom of the Shi'ite saint Hussein in 680 (Ashura) down to the twentieth century. Even more remarkable is the eighteen- or nineteen-century mnemonic gap we see in most national calendars throughout the Christian world—an official commemorative blackout that usually begins right after the assumption of Mary in the first century and is ultimately broken off only by the glow of relatively modern sociomnemonic beacons such as the British settlement of Australia in 1788 (Australia Day), the storming of the Bastille in 1789 (Bastille Day), or the American Revolution (Fourth of July).

Indeed, of the 191 national calendars I have examined, only twenty-two actually invoke the memory of any specific historical event that happened (other than the celebrated European "discovery" of America in 1492), or figure who flourished, between 680 and 1776; and in thirteen of those twenty-two cases, only the sixteenth or seventeenth century is involved. Thus, around the entire globe, only nine countries actually commemorate on their national holidays anything specifically related to the period from 680 to 1492: India (the birth of Guru Nanak, the founder of Sikhism, circa 1469), Hungary (the reign of King Stephen I, from 1001 to 1038), the Czech Republic (the birth of Slavonic culture in 863 and the martyrdom of Jan Hus in 1415), Lithuania (the coronation of Grand Duke Mindaugas circa 1240), Andorra (the joint suzerainty agreement between France and the bishop of Urgel in 1278), Slovakia (the birth of Slavonic culture in 863), Switzerland (the establishment of the Swiss confederation in 1291), Bulgaria (the invention of the Cyrillic alphabet in 855), and Spain (the reputed discovery of Saint James's body in Compostela in 899). This also means that, at least as far as calendrical commemoration is concerned, the eighth, tenth, twelfth, and fourteenth centuries are considered virtually "empty" worldwide!

Needless to say, such seemingly barren historical valleys were never really empty. *In 1926,* an intriguing, richly textured portrait of a typical ordinary (and therefore "unmarked") year, effectively demonstrates that even periods that we may later come to recall as practically empty were in fact quite eventful,[101] thereby reminding us of the fundamental difference between history as it actually occurs and the way it is conventionally remembered.

Legato and Staccato

Regardless of the specific form of historical narrative we use to help us impose some retrospective structure on the past, there are two basic modes of envisioning the actual progression of time within it. While one of them features essentially contiguous stretches of history smoothly flowing into one another like the successive musical notes that form *legato* phrases, the other tends to highlight unmistakably discontinuous breaks separating one seemingly discrete historical episode from the next, like the successive notes that form *staccato* phrases.[102] As we can see in figure 10, whereas in the first type of *historical phrasing* change is basically viewed as *gradual,* as manifested in the way we tend to narrate the unmistakably *continuous* progression of one's skills as a reader or chess player, in the second, by contrast, it is quite *abrupt,* as manifested in the way we normally narrate medical or military careers.[103] As exemplified by the way we use concepts like "style" and "wave" when narrating the histories of art and immigration, these two general modes of envisioning change entail two rather distinct visions of the past. Nowhere, however, is the fundamental contrast between those visions sharper than in the way we narrate the history of life on this planet.

Gradualist paleobiological narratives[104] are essentially a temporalized form of the classical image of natural plenitude commonly known as the Great Chain of Being.[105] Exemplifying such a narrative is Darwin's theory of organic evolution, which does not recognize any "leaps" in nature and basically envisions species mutating by short, slow steps.[106] Evolution is thus a gradual process in which a perfectly graded chain of intermediate forms evolve from one another almost imperceptibly, with no sharp cut-

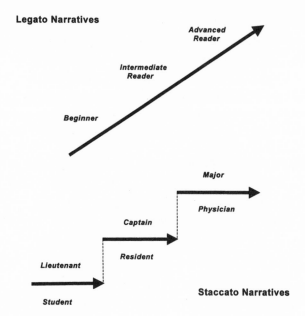

Figure 10 *Historical Phrasing*

offs.[107] Any breaks between species are therefore only an illusion resulting from an imperfect fossil record. If it were perfect, an insensibly graded fossil sequence formed by every possible transitional "missing link" connecting successive species would allow us to actually see the unmistakably continuous nature of biotic evolution.[108]

Staccato paleontological narratives, by contrast, consist of discrete historical episodes separated from one another by pronounced breaks marking abrupt, rapid changes. Both Georges Cuvier's and Louis Agassiz's catastrophist visions of history as essentially punctuated by dramatic climatic upheavals are perfect examples of such narratives. So is Niles Eldredge and Stephen J. Gould's punctuated equilibrium scenario, which features episodes of rapid speciation involving sharp, sudden interspecific breaks.[109] Species are thus envisioned as occupying discrete historical niches, with breaks in the fossil record essentially reflecting the actual biological gaps between them. "Missing links" clearly have no place in such a narrative.[110]

Each of these general visions of change represents ¿
monic tradition often associated with a specific commun
professional socialization, different generations of biol‹
come to envision the past as members of altogether di
communities. While gradualism was the predominant
ing the history of life for more than a century after D
equilibrium theory has been the commonly accepted way of doing it for
the past twenty-five years.

As we shall see in the next two chapters, "legato" narratives are naturally
quite indispensable to any effort to establish historical continuity. However,
as we shall see in the last two chapters, "staccato" narratives are inevitably
at the heart of any attempt to introduce some historical *dis*continuity. As
we try to organize the past in our minds, we clearly seem to need, and in
fact frequently use, both.

Historical Continuity

Not every historical narrative necessarily presupposes change. Essentially projecting a general sense that there is "nothing new under the sun," many, in fact, regard the present as a *continuation* of the past.[1] Thus, instead of one actually replacing the other,[2] the two are viewed as parts of an integrated whole.

Despite the conventional grammatical distinction between the past and present tenses, *the past and the present are not entirely separate entities.* The notion that we could actually identify a point prior to which everything is "then" and subsequent to which everything is "now" is an illusion. So is the idea that we can somehow determine unequivocally how many years must pass before we can actually feature something in a history textbook or a "historical" museum.

The ways in which we organize our diet, interpersonal etiquette, and personal hygiene are essentially habitual patterns continually *perpetuated* as part of a social *tradition.*[3] In a similar vein, we still use words that were around in the fourteenth century, and our scientists typically frame their current investigational agendas in terms of formal expectations (hypotheses) that are based on past research. As so clearly manifested in the ubiquitous role of precedent in common law, the present is largely a *cumulative,* multilayered collage of past residues continually deposited through the cultural equivalent of the geological process of sedimentation.[4]

Social relations, too, are historically embedded, as demonstrated by the great difficulty so many Romeo Montagues and Juliet Capulets seem to have in extricating their current (not to mention potential) ties from the ever-present grip of their ancestral pasts, thereby substantiating Karl Marx's observation that "the tradition of all the dead generations weighs like a nightmare on the brain of the living."[5] As German president Roman Herzog reminded Poles on the fiftieth anniversary of the 1944 Warsaw uprising against the Nazi occupation, "only history divides us now."[6] Yet such divisions are not so easy to erase. As one Irish-American commentator shrewdly observed after being chastised by her mother for planning to stay at a hotel named after Oliver Cromwell, who is still loathed by Irish nationalists, "in Irish time, 1651 and 1981 were only moments apart"![7] And as the director of the Oñate Monument and Visitors Center pleaded when Native American militants attempted to saw off the foot of a bronze statue of Juan de Oñate, the brutal Spanish conquistador who cut off the feet of those who resisted his conquest of New Mexico in 1599: "Give me a break—it was 400 years ago. It's OK to hold a grudge, but for 400 years?"[8]

Given all this, ignoring the historical background of present situations is somewhat analogous to living in a two-dimensional Flatland.[9] Regarding such situations as if they have no past is like a physician failing to ask a patient about diseases that run in her family. It basically puts one in the situation of a child who is just starting to read newspapers and is still unfamiliar with the tacit historical background of the stories he reads, so it is virtually impossible for him to understand them fully. Such quasi-amnesic dissociation of current events from their historical contexts is therefore tantamount to chopping up a film into seemingly disconnected stills.

As exemplified by ex-convicts' difficulties in finding a job after serving their prison sentence and the fact that former nuns often wear particularly provocative clothes and makeup to avoid being perceived as "sweet" and naïve,[10] the past is also considered an integral part of present identities. That explains the identity crises we often experience as a result of dramatic changes that quite literally tear us from our past, as when we emigrate, undergo a hysterectomy, or lose a spouse.

Yet the continuity between the past and the present is also disrupted nowadays by the tremendous acceleration of social and technological change[11] and the rise of a distinctly modern economy based on disposability and planned obsolescence.[12] That, however, has triggered an unmistakably *conservative* urge to preserve such continuity (and a corresponding strong aversion to any change that might threaten our identity), as manifested in various traditionalist efforts (such as by the Amish) to conserve the old ways of life as well as in high-school yearbooks, oldies radio stations, and numerous other expressions of nostalgia.[13]

Predictably, we feel particularly nostalgic about those parts of our past that seem most hopelessly irrecoverable. (As I leaf through a collection of mementos from the 1950s,[14] it is old matchboxes, chewing-gum wrappers, and magazine covers that most evocatively touch the child in me.) We likewise experience nostalgia during periods of dramatic change. It is upon leaving home to go to college that we often become sentimentally reattached to our childhood belongings, and upon retiring that we suddenly long for our lost youth.[15]

As demonstrated by the wave of nostalgia that swept the United States in the late 1970s as people began to grasp the full scope of the tremendous social changes that had taken place around them since the 1960s,[16] that sentimental reaction applies to groups as well as individuals. Yearning for yesterday is particularly pronounced when a group experiences a sharp political, cultural, or economic downturn, as exemplified by the sentimental longings of nineteenth-century Arab historians witnessing the beginnings of Europe's colonial expansion and the decline of the Ottoman Empire to the past glories of medieval, Muslim Spain.[17]

As an attempt to reconnect with older layers of oneself and thereby gain access to some long-gone past, nostalgia inevitably raises the philosophical question of how identities can indeed *persist* in the face of constant change. We certainly cannot accept such persistence as a given. After all, not a single cell in my body was there forty years ago and not a single current member of "the French nation" was alive during the French Revolution. Yet despite the fact that Geoffrey Chaucer would most probably have difficulty following a conversation among American teenagers today, we nevertheless do regard "the English language" as an entity that has persisted continually throughout the past six centuries. By the same token, we regard Italy's national soccer team (the *Azzurri*) and Harvard University's psychology department as essentially *uninterrupted* entities, although their memberships obviously keep shifting.[18]

Yet how do we actually overcome the fact that the fourteenth- and twenty-first-century versions of what we consider "the same" language or social group are not really contiguous? How do we actually manage to establish *historical continuity* between virtually *non*contiguous points

in time, thereby essentially transforming an assemblage of utterly disconnected "successive perceptions"[19] into a seemingly coherent, *constant*
identity?

As we shall see, such seemingly self-evident constancy is only a figment
of our minds.[20] As David Hume so rightly pointed out, it is not really a quality of objects but of the way we perceive those objects.[21] Continuous identities are thus products of the *mental* integration of otherwise disconnected
points in time into a seemingly single historical whole. More specifically,
it is our memory that makes such mental integration possible, thereby al-
lowing us to establish the distinctly mnemonic illusion of continuity.[22] As
victims of Alzheimer's disease and other forms of memory loss make so
painfully clear, maintaining a continuous identity is virtually impossible
without the essentially "adhesive" act of memory.

The various mnemonic strategies we use to help us create the illusion
of historical continuity typically involve some mental *bridging*. A prototypi-
cal facilitator of integrating noncontiguous spaces,[23] the bridge is a perfect
metaphor for the mnemonic effort to integrate temporally noncontiguous
manifestations of what we nevertheless consider "the same" entity (person, organization, nation). And in the same manner that we try to "bridge"
the historical gap between the present and the future through the use of
conventional "adhesive" farewell clichés such as "I'll see you later" (or its
many cross-cultural functional cousins, such as the Italian *arrivederci* and
the German *auf wiedersehen*),[24] we also use various mental bridging techniques to produce the "connecting historical tissue"[25] that helps us fill any
historical gaps between the past and the present.

These techniques typically involve some mental editing to produce an
illusory *quasi-contiguity* that can help offset the actual temporal gaps between noncontiguous points in history. Like the pasting we do in word
processing, such editing resembles cinematic montage, in which a series
of altogether separate shots are essentially pasted together to form a single,
seemingly seamless film.[26] As we can see in figure 11, such *mnemonic pasting* helps us mentally transform series of noncontiguous points in time
into seemingly unbroken historical continua.

Same Place

Despite the fact that mnemonic bridging is basically a mental act, we often
try to ground it in some tangible reality. Indeed, one of the most effective

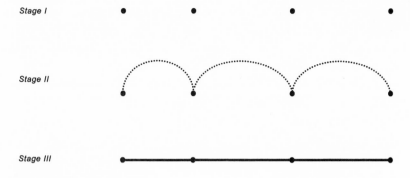

Figure 11 *Mnemonic Pasting*

ways of bridging the gap between noncontiguous points in history is by establishing a connection that allows them to almost literally touch one another.

Constancy of place is a formidable basis for establishing a strong sense of sameness. Even as we ourselves undergo dramatic changes both individually and collectively, our physical surroundings usually remain relatively stable. As a result, they constitute a reliable locus of memories and often serve as major foci of personal as well as group nostalgia.[27] In providing us with some sense of *permanence,* they help promote the highly reassuring conservative illusion that nothing fundamental has really changed.

That explains why, at their national convention in Los Angeles in 2000, Democrats kept reminding Americans that it was *at that same place* that they had nominated John Kennedy as their ultimately victorious presidential candidate forty years earlier. It also underscores efforts to literally *preserve* the past in historic buildings and neighborhoods.[28] Whether in Stockholm, Granada, or Montreal, the "impulse to preserve" the past is "a reaction against the increasing evanescence of things and the speed with which we pass them by."[29]

Such obvious concern about historical continuity is also the reason why, after conquering Córdoba in 1236, King Ferdinand III of Castile did not destroy the gorgeous mosque that for 450 years had epitomized the splendor of Moorish architecture. Instead, he converted it to a cathedral (which, nevertheless, is still known eight centuries later as the Mezquita), thereby integrating Spain's Muslim past and Christian present in a most visually compelling manner. An equally evocative spectacle awaits anyone who enters Hagia Sophia, the great church that was built by Emperor Justinian in 537 and converted by Sultan Mehmet II in 1453 to a mosque and again by President Kemal Atatürk five centuries later to a state museum. The visual fusion of Istanbul's Byzantine and Ottoman pasts with its modern Turkish present in the same building is a spectacular sight, not to mention a most remarkable instance of mnemonic engineering.

Constancy of place also allows us to virtually "see" the people who once occupied the space we now do.[30] As we look into the eerily empty kitchen of a fully preserved house in Pompeii, we can quite vividly visualize a family working there at the very moment Mount Vesuvius erupted nineteen centuries ago, enabling us to actually identify with those people. Walking down the streets of an old city, we can "make contact with previous generations" by literally walking in their footsteps and looking at the "vistas that greeted their eyes."[31] (Standing outside Marco Polo's house in Venice and looking at what he could see from his window 750 years ago, I could actually feel the overwhelming sense of claustrophobia that must have stricken the man most responsible for expanding medieval Europe's geographical horizons.) Such identification is often exploited by authors of "then and now" mental time-travel books who superimpose transparent overlays of imaginary scenes from antiquity onto actual photographs of historical ruins in their present-day surroundings.[32] It also explains the great touristic appeal of old inns where George Washington allegedly spent a night more than two centuries ago.

Indeed, place plays a major role in identity rhetoric. For example, whether it involves devout Muslims going to Mecca on their hajj, patriotic Americans coming to Philadelphia to see the Liberty Bell, or romantic couples revisiting the site of their first date, *pilgrimage* is specifically designed to bring mnemonic communities into closer "contact" with their collective past. This mnemonically evocative aspect of place likewise underscores the role of *ruins* in solidifying such ties.[33] Thus, during the Russian bombing of their region in 1999, Chechens felt highly protective about the old stone towers that for many centuries helped "connect" them to their ancestors.[34] This evocative aspect of place also explains the tremendous significance of the archaeological excavations at Masada for modern Israeli nationalism.[35] Having its young soldiers take their oath on that mountaintop has certainly helped Israel claim the legacy of the ancient Jewish warriors who died there nineteen centuries ago.[36] Similar concerns about historical continuity led the shah of Iran in 1971 to stage the public commemoration of the 2,500th anniversary of the foundation of the Persian Empire among the ruins of the ancient city of Persepolis.[37]

The relation between place and identity has unmistakably essentialist overtones. Thus, to early Egyptian nationalists, ancient and modern Egyptians "were *inevitably* subject to identical . . . influences,"[38] having both lived in the Nile Valley. A highly romanticized "natural" link between geography and nationhood likewise underscores the special significance of Zion to the modern political movement that basically derives its entire public identity from its name. For Zionism, Palestine's physical landscape literally bridges the 1,800-year historical gap separating its ancient and modern inhabitants.[39] That is why modern ultranationalist Jewish settlers on the West Bank ("Judaea and Samaria") are so strongly attached to their settlements. As one settler in the old city of Hebron explains,

> You feel here such a deep *connection*. On this mountain stood the palace of Kind David. Here, *right here*, God promised Abraham the Land of Israel. . . . Just imagine to yourself that *I go to sleep at the very place where Abraham used to get up every morning*! . . . What Jew wouldn't want to *live near Abraham*?[40]

Relics and Memorabilia

Yet mnemonic "connectedness" need not depend on constancy of place. After all, even strictly physical mnemonic bridges can be detached from actual places, as exemplified by souvenirs, mementos, and other *memorabilia*. Despite the fact that they are not tied to a specific location, the actual material essence of such portable *relics* helps provide some physical continuity, which is why they are indeed used almost exclusively, as their etymology suggests, for storing memories. Like stuffed animals, security blankets, and other "transitional objects" used by infants as highly effective existential bridges,[41] relics basically allow us to live in the present while at the same time literally "cling" to the past.

Furthermore, the fact that they are not tied to a particular location certainly allows much more flexibility in the way we use such "reminders." Unlike old neighborhoods, for example, the portable nature of relics means that they can help us recall past events without our having to be physically present at the place where they actually occurred. (Lying in a hammock that is now hanging in my backyard in New Jersey instantly evokes vivid memories of cuddling my son in the same hammock fifteen years ago in my old backyard on Long Island.) Like a faraway lover's letter or lock of hair, such objects allow us mnemonic access even to persons and places that are no longer physically accessible—hence the tremendous importance of Torah scrolls to Jewish communities in exile throughout history, and the special significance of home-related memorabilia as *tangible* links between past and present selves when we go to college.[42]

Given the role of memorabilia as a mnemonic bridge, we often refuse to throw out old clothes and hold on to otherwise worthless presents we received from people who once occupied a special place in our lives. As we keep moving through life from one place to another, the various mementos we carry with us make it, somehow, much easier to maintain the continuity between our past and present selves. Because they can no longer cling to their past personal belongings, the psychological recovery of war refugees as well as earthquake, flood, and hurricane survivors is particularly difficult.[43]

Like ruins and historic buildings, relics and memorabilia offer us a remarkably vivid, quasi-tangible contact with the past. I recall the exhilarating experiences of looking at one of the original copies of the 1455 Gutenberg Bible at the British Museum and standing, at the Topkapi Saray Museum in Istanbul, in front of a sandal reputedly worn by Mohammed fourteen centuries ago! (Displaying to his followers from the top of a mosque in Kandahar an old cloak allegedly worn by the Prophet in fact helped Mullah Omar launch the Taliban's highly traditionalistic Islamic revolution in Afghanistan in 1996.)[44] Our appreciation of such "tangible" contacts with the past explains our tremendous fascination with the Scala Santa (a Roman staircase reputedly made from the actual steps that Jesus climbed when he was brought before Pontius Pilate) and the Shroud of Turin[45]—not to mention the legendary Holy Grail. It also explains why we keep scrapbooks and the significant role of museums in promoting nationalism.[46]

It is precisely their evocative function that makes relics such as the American flag that was in Abraham Lincoln's theater box the night he was assassinated[47] so valuable. Why else, after all, would anyone be willing to pay so much for a 1957 Cadillac or a half-broken manual typewriter? Only their role in providing some sentimental connection to the past makes such *antiques* so precious to us.[48]

As anyone who has ever been asked to pose for a picture or sign a guest book very well knows, we are more than just passive consumers of memorabilia. As exemplified by the plaques, medals, award certificates, and other purely commemorative objects we produce, we often actively design such future sites of memory well in advance! Like school yearbooks (for which class pictures are nowadays sometimes taken before the actual school year begins), the tremendous value of such "pre-ruins"[49] lies in their being highly evocative and thus able to constitute quasi-tangible bridges to future pasts.

Imitation and Replication

Along with trying to approximate actual physical contact between the past and the present, we also try to generate various iconic representations[50] of the past that would at least *resemble* it. Consider, for example, the replicas of King Nebuchadnezzar's monumental buildings in Babylon constructed by Iraqi president Saddam Hussein almost 2,600 years later.[51] Consider also, in this regard, nineteenth-century experiments with neoclassicism,[52] and various attempts by U.S. colleges to project an "old" look through the use of neo-Gothic architecture.

The physical resemblance between the images we try to capture in statues and portraits and the actual persons they are designed to later invoke represents similar attempts to somehow compensate for the lack of actual physical contact between the past and the present. Such *iconic connectedness* is even more spectacularly evident in the remarkably vivid images of those persons that we try to capture in photographs, let alone on video or film.[53]

Our attempts to *imitate* the past and thereby "reproduce" it are also expressed through our appearance and behavior. Indeed, much of what we call "tradition" consists of various ritualized efforts to become more fully integrated into our collective past through imitation. The remarkable preservation of many archaic behavior patterns is evident in religious ritual, courtroom etiquette, parliamentary procedure, military drills, folk dances, and ethnic cuisine. It also accounts for the unmistakably traditionalistic ceremonial garments of kings, popes, graduating classes, and national soccer teams.

Even more spectacular in this regard are historical *revivals* such as the restoration of the ancient Roman salute by the Fascists in Italy, the resurgence of Hebrew as an everyday language in modern Israel, and various "invented traditions."[54] By generating new traditions that nevertheless seem old (such as Kwanzaa, a pseudo-African festival essentially invented in California in the 1960s),[55] such revivals are designed to create the illusion of historical continuity since time immemorial. As exemplified by the invention of the Highland "tradition" in Scotland two centuries ago[56] or the relatively recent adoption of traditional African names and garb by American black nationalists, however, such continuity is a mere figment of our minds.

Imitating entails repetition,[57] thereby helping to create an illusion of actual *replication*. By wearing clothes resembling those worn by our ancestors and eating the "same" food they once ate, we try to symbolically relive their lives. Such simulative attempts to "relive" the past are particularly evident in ritual pageants involving actual *reenactment*, such as the 1995 event that literally "retraced the steps" of the 1965 civil rights march from Selma to Montgomery (this time, however, featuring the repentant former Alabama governor George Wallace singing "We Shall Overcome").[58] The common use of period costume on such occasions helps maintain the illusory conflation of the present and the past. Such *quasi-synchronicity* is further enhanced through constancy of place, as in Colonial Williamsburg and other "living history" museums, where quasi-authentic guides use the present tense when talking with visitors about the eighteenth century! It can nowadays also be produced digitally, as demonstrated by Natalie Cole in her stunning 1991 recording of "Unforgettable." In it she sings "along with" her father, Nat King Cole, who had already been dead for a quarter of a century.[59]

"Same" Time

Historical reenactments often take place at Christmas, Thanksgiving,[60] and other holidays. (Having grown up in Israel, I have vivid childhood memories of "coming out of Egypt with all my belongings" on Passover and "bringing my first fruits to the ancient Temple in Jerusalem" on Shavuot.) Indeed, *periodic fusion with the past* is the very essence of annual (birthdays, holidays) and other (silver weddings, high-school reunions, bicentennials) *anniversaries*.[61] And this fusion is even more evocative when synchrony is combined with constancy of place, as in the annual "replay" in Nazi Germany of the 1923 Munich Beer Hall Putsch at the same place as well as on the "same" day (9 November),[62] or the peace rallies held every year in Israel on 4 November at the site of the 1995 assassination of Prime Minister Yitzhak Rabin.[63]

Solidifying such periodic fusion with the past through the establishment of an *annual cycle of commemorative holidays* is one of the main functions of the calendar.[64] (In helping ensure that we periodically "revisit" our collective past, the calendar also plays a major role in our mnemonic socialization. For instance, long before Americans are formally introduced in school to the English colonization of their country, they learn through their annual observance of Thanksgiving about the seventeenth-century Pilgrims who settled New England.)[65] And despite the difficulty of compressing thousands of years of history into a 365-day holiday cycle, we nevertheless try to combine our linear and circular visions of time in an effort to somehow "synchronize" our annual holidays with the historical events whose memory they are designed to evoke. Thus, when Jews bless God on Hanukkah for the miracles he performed *"in those days at this time,"* they are simultaneously associating that holiday with a particular time in history (the Maccabean Revolt of 165 BC) as well as a particular time of year (the end of the month of Kislev).[66] Such symbolic *synchrony of "now" and "then"* reflects our conservative urge to do away with the very distinction between them.

Most "holy days" are symbolically associated, and therefore also calendrically "synchronized," with certain *days* in a group's history: Malta's Victory Day, with the lifting of a four-month Ottoman siege on 8 September 1565; New Zealand's Waitangi Day, with the signing of the celebrated treaty between the islands' Maori and British populations on 6 February 1840; Colombia's Battle of Boyacá Day, with Simón Bolívar's victory over Spain on 7 August 1819; and so on. Yet even such remarkable effort to literally *synchronize calendrical and historical time* certainly pales compared to the church's unparalleled sociomnemonic accomplishment of featuring the three calendar months from Ash Wednesday to Pentecost as a perfect calendrical replica of three specific historical months in the year AD 30![67]

As one might expect, synchrony of this sort has unmistakably essentialist connotations given the exceptionally evocative *seasonal identity* of the historical "then" and the calendrical "now." Whereas eating the "same" unleavened bread on Passover helps present-day Jews identify with the ancient Israelites who allegedly came out of Egypt three thousand years ago, the fact that it takes place at the same time of year as the Exodus is specifically designed to make the link between them seem more "natural."

There is absolutely nothing natural, however, about annual anniversaries. Essentially using 260- and 210-day holiday cycles,[68] neither Guatemalans nor Indonesians, for example, evidently tie their traditional notions of the "same" time to the seasons. Various memorial services held on 11

December 2001 to mark the three-month "anniversary" of the 11 September attack on the World Trade Center likewise remind us that only social convention ties birthdays and other holidays to the annual revolution of the earth around the sun.[69]

Highly cognizant of the mnemonic role of anniversaries, we often schedule special events for particular dates that are already imbued with historical significance. It was by no means a merely random coincidence (but, rather, a deliberate *calendrical coincidence*) that Mexico's 1917 constitution, for example, was promulgated on the "same" day as its 1857 precursor (5 February), or that the upper house of Denmark's parliament was abolished in 1953 on the "same" day marking the end of the absolute monarchy there 104 years earlier (5 June). Similar sociomnemonic sensitivities must have played a major role in Saddam Hussein's decision to assume the presidency of Iraq on the eleventh anniversary of the 17 July coup that brought his Ba'ath Party to power in 1968, as well as in Hungary's decision in 1989 to proclaim its post-Communist republic on the anniversary (23 October) of its historic anti-Soviet uprising in 1956. And when Timothy McVeigh bombed the Murrah Federal Building in Oklahoma City on 19 April 1995, his attack was designed to mark the second anniversary of the destruction by government agents of the Branch Davidian cult compound in Waco, Texas, which he evidently wished to avenge.

Historical Analogy

The aforementioned cases of Denmark, Hungary, and Mexico also underscore our tendency to view the past as somehow "similar" to the present, thus linking them through *analogy*.[70] That happens, for example, when we unwittingly "reproduce" early patterns of relating to our parents in the way we now relate to authority figures or choose our sexual partners. It is often also done quite consciously. The famous affair between Camilla Parker Bowles and Britain's Prince Charles was in fact initiated by her telling him that her great-grandmother and his great-great-grandfather had been lovers![71]

The tendency to invoke the past analogically, then, characterizes more than just lawyers in search of judicial precedents.[72] The totally unanticipated destruction of the World Trade Center, for example, was immediately compared by many to the Japanese attack on Pearl Harbor sixty years earlier, just as the political situation in Bulgaria in 1945 was described at the time as a Bulgarian "version" of the situation in Russia in 1917,[73] and as Yitzhak Rabin and Ehud Barak were compared by critics of their conciliatory approach toward the Palestinians to France's Marshal Henri Pétain, another "former military hero who as a political leader later betrayed his country." (Barak was also compared by ultranationalist Israeli rabbis to the spies who were sent by Moses to reconnoiter Palestine, only to be intimidated by its native population.)[74] And in the same way that John F. Kennedy's decision to increase U.S. military aid to South Vietnam was inspired by the successes of "analogous" campaigns against Communist insurgencies in Malaya and the Philippines,[75] it was the humiliating memory of "the ignominy of Muslims being driven out of Europe by Christian armies in the 15th century"[76] that so evocatively affected arch terrorist Ayman al-Zawahiri's view of Islam's present relations with the West. Such pronouncedly anachronistic mental fusions of past and present also explain the tremendous significance of the Exodus to both English Puritans in the 1640s and American colonists in the 1770s.[77]

Like generals preparing for the last war, we often draw on analogous ("similar," "parallel") situations from the past when facing current ones. Incorporating the "lessons" of the 1815 Congress of Vienna was an integral part of drafting the 1919 Treaty of Versailles.[78] Mobilizing memories in this manner usually involves efforts to avoid past mistakes, such as behavior for which others have already been penalized.[79] Indeed, we often use past traumas as a scare tactic, as exemplified by the strategic manipulation by the U.S. government of the memory of the 1918 influenza epidemic to promote mass immunization during the 1976 swine flu scare,[80] and of the 1936 Berlin Olympics by advocates of boycotting the 1980 Moscow Olympics following the Soviet invasion of Afghanistan (as well as by opponents of Beijing's bid to host the 2008 Games),[81] not to mention the post-Holocaust "Never again" rhetoric of the Jewish Defense League.

Worries about "repeating" past mistakes were also evident in America's fear of repeating in Vietnam the "loss" of China to the Communists in 1949[82] as well as in its efforts to avoid in 1945 the "same" mistakes made with regard to Germany at the end of World War I.[83] And when Congress gave the president its almost unanimous support after the 2001 attacks on the World Trade Center and the Pentagon, Senator (and Vietnam veteran) John McCain explicitly cautioned against "repeating" the mistakes made after the 1964 North Vietnamese attacks in the Gulf of Tonkin, which helped drag the United States into a long, unwinnable war.[84] The wish to avoid a "repeat" of the Cultural Revolution and the Ruby Ridge and Waco debacles also affected the way Deng Xiaoping and U.S. law enforcement officials handled the 1989 student demonstrations in Tiananmen Square[85] and the 1996 standoff with the antigovernment extremists Freemen of Montana, respectively.[86]

Consider also the infamous British and French attempt to appease and thereby "contain" Adolf Hitler by practically sacrificing Czechoslovakia at Munich in 1938. The memory of the tragic consequences of that inglorious event was already invoked in 1945 when the United States was cautioned by Iran not to let the Soviet Union do in Azerbaijan what Germany had done to Czechoslovakia in 1939.[87] Five years later, when North Korea invaded South Korea, it likewise played a major role in U.S. president Harry Truman's decision to help the South[88] as well as in the ensuing development of the "domino theory" that helped shape U.S. foreign policy in the 1950s and 1960s.[89] The "lesson" of Munich also played a significant role in Britain's decision to attack Egypt when Gamal Abdul Nasser seized the Suez Canal in 1956,[90] and explicit comparisons of Yassir Arafat and Slobodan Miloševic to Hitler were still used in 1982 and 1999 to justify the Israeli invasion of Lebanon and the bombing of Serbia by NATO. Warning the world not to "repeat" its regrettable capitulation to the German dictator by sacrificing Kuwait also helped U.S. president George Bush mobilize international support for the Gulf War.[91]

Like any other symbol, historical analogies clearly transcend their historical specificity. When drawing such analogies we therefore do not feel constrained by the considerable temporal distance often separating past signifiers from their corresponding present signifieds. Their evocative power is much greater, however, when the cultural affinity between the two helps offset such distance, as exemplified by the use of the Persian army defeated by Alexander the Great in 333 BC to allegorically represent in a 1529 painting the Ottoman troops laying siege to Vienna that year;[92] the symbolic identification of Peruvian president Alejandro Toledo with the fifteenth-century Inca emperor Pacachutec;[93] or the Nazi portrayal of the Roman wars against Carthage as a racial conflict between Aryans and Semites![94] Just as evocative in this regard were Sergei Eisenstein's famous 1938 cinematic tribute to Prince Alexander Nevsky, who in 1242 thwarted a German invasion of Russia; the depiction of Jewish troops preparing to confront the German army about to enter Palestine in 1942 as a "new edition" of the defenders of Masada;[95] and the popular portrayal of Arafat and Osama bin Laden as modern versions of Saladin, the highly revered architect of Islam's historic victory over the Crusaders in 1187.[96]

The celebrated victory of this twelfth-century Mesopotamian warrior was similarly featured by Saddam Hussein right before the Gulf War as prefiguratively suggestive of the outcome of (and therefore also a very useful *model* for) his own impending battle against the latter-day invading infidels from the West. An exceptionally shrewd manipulator of such cultural and geographical "parallels," he had portrayed himself a few years earlier during the Iran-Iraq War as a present-day incarnation of Sa'd ibn-abi-Waqqas, the Arab general who defeated the Persians at Qadisiya in 637, and even issued colorful postage stamps anachronistically commemorating "Saddam's Battle of Qadisiya."[97] Calling for a new Arab war against Israel, he then proceeded to link himself analogically to Nebuchadnezzar II, the celebrated Babylonian (and, as such, "Iraqi") king who managed to conquer Jerusalem and destroy the ancient Israelites' First Temple in 586 BC.

Throughout their 1996 standoff with the Freemen of Montana, U.S. law enforcement agents were explicitly trying to avoid "*another* Waco."[98] By the same token, when Israeli prime minister Ariel Sharon worried that the United States might be willing to compromise Israel's security to buy Arab support for its war against terror, he made it unequivocally clear that Israel "will not be Czechoslovakia."[99] In a somewhat similar vein, when President Lyndon Johnson wanted to avoid public criticism by General William Westmoreland during the Vietnam War, he invoked a "parallel" incident from the Korean War involving Truman and General Douglas MacArthur and explicitly warned him not to "pull a MacArthur on me."[100] On the eve of the Battle of Khe Sanh, invoking the historic French defeat at Dien Bien Phu fourteen years earlier, he likewise warned that he did not want "any damn Dinbinphoo."[101]

As demonstrated by the above—and by the use of *Purim*, the name of a traditional commemoration of the alleged deliverance of Persian Jewry from a massacre plotted twenty-four centuries ago, to denote any "such" event involving Jewish communities throughout history[102]—the events to which we make historical analogies are basically regarded as transhistorical, generic symbols. Based on a perceived similarity between "parallel" situations, such analogies thus clearly presuppose some mnemonic typification. That explains how American colonists could view their bondage to England as "a second Egypt"[103] and how the U.S. interventions in Vietnam and the Dominican Republic were actually designed to prevent "another"

Korea or Cuba.[104] It also explains the wish to avoid any more "Munichs" or "Vietnams."[105]

Discursive Continuity

Like holidays and other anniversaries, historical analogies underscore the fact that our "ties" to the past are not always physical or even iconic but quite often purely symbolic. That is certainly true of the ties between noncontemporary namesakes. The tremendous mnemonic significance of names as discursive tokens of "sameness" helps explain, for example, why the rebels in Chiapas would choose to adopt the name *Zapatistas* more than seventy years after the actual death of the revered hero of the Mexican Revolution, Emiliano Zapata.[106] This wish to establish such seemingly direct "links" to the past likewise led post-Communist Mongolian nationalists to name a new vodka after their thirteenth-century national hero, Genghis Khan.[107]

A somewhat similar discursive form of bridging the historical gap between the past and the present is the subtle use of *consecutive ordinal numbers* to imply temporal contiguity. The so-called Third Reich, for example, was thus featured by the Nazis as a *direct* successor to the "second" (1871–1918) German empire, thereby tacitly glossing over the pronouncedly nonimperial fifteen-year period actually separating them (not unlike the forty-four-year period from 1804 to 1848 separating France's so-called First Republic from the Second). The name *Menelik II* was similarly designed to help Ethiopians spin a mental thread "linking" their late nineteenth-century emperor to the legendary founder of their kingdom despite the fact that 2,800 years separated their reigns (just like the name linking twentieth-century Bulgarian czar Boris III to his tenth-century namesake, Boris II). Building their dream "Third Temple" in Jerusalem would likewise help Jewish ultranationalists dim the memory of the nineteen-century nationalist "void" that began with the Roman destruction of the Second Temple in AD 70.

Another discursive form of "bridging" historical gaps is the use of a *single continuous timeline* for chronological dating.[108] In marked contrast to episodic "eras" tied to the inevitably discrete reign of a specific monarch, for example,[109] the standard Jewish, Christian, and Muslim eras feature continuous timelines that can actually "link" any given points in history! Like poems that forgo conventional spaces between words or books that end with thirty-six-page passages virtually uninterrupted by even a single comma,[110] such timelines embody an unequivocal commitment to continuity. Such commitment has been at the heart of the feminist critique of the conventional segmentation of women's lives into the supposedly discontinuous biographical phases associated with the social titles *Miss* and *Mrs.*, leading, indeed, to the introduction of the single (and thus essentially "continuous") title *Ms.*[111] It has likewise inspired nationalist attempts to challenge the conventional periodization of Egyptian history in accordance with its various conquests by foreigners. By essentially "Egyptianizing" its Nubian, Persian, Roman, Arab, and Mongol conquerors and presenting the (Macedonian) Ptolemys and (Turkish) Mamluks as full-fledged Egyptian monarchs, Egypt was thus portrayed as having basically remained "the same" throughout its five-thousand-year history.[112]

As one might expect, this portrayal also greatly resembles the way individuals normally produce a *continuous biography,* which, as job interviews, high-school reunions, and other "autobiographical occasions"[113] can attest, is a considerable discursive accomplishment that cannot ever be taken as a given.[114] The reason it takes such an effort to revise an old résumé is not only because so much has happened in our life since the last revision but also because of our obvious need to keep "updating" our past so as to make it congruous with our often-changing present self-image. Whether they are made by the overweight, prematurely aging alcoholic who was once considered the most popular girl in her class or by the former prankster who is now a prominent judge, any attempts to discursively "align" our past and present underscore our overall wish to present to the world an essentially *continuous self.* It is the social unacceptability of any major biographical incongruities between past and present identities that makes blackmailing such a lucrative business.[115]

The discursive production of a continuous biography consists of playing up those elements of our past that are consistent with (or can somehow be construed as prefiguring) our present identity while downplaying those that are incongruous with it. That process entails invoking the classic Aristotelean distinction between the "essential" aspects of an object that we believe constitute its "true" identity and those we conventionally consider merely "accidental." Whereas my driver's license and social security number are specifically designed to confirm that I am still "the same" person even if I lose 56 percent of my body weight through bariatric surgery,[116] which socks I am wearing today or how much milk I take with my coffee are not considered part of my "essence."[117] (By the same token, unless the actual piece on which its official Vehicle Identification Number is engraved has been removed, a car whose engine and four doors have all been replaced is still considered to be "the same" car.)[118] In fact, as exemplified by self-accounts of past marriages or periods of clinical depression, in order to downplay biographical incongruities between past and present selves, we sometimes dismiss even lengthy stretches of action (philandering) or inaction ("vegetating") as somehow uncharacteristic of who we "actually" are.[119] In order to produce a seemingly continuous female biography, a male-to-female transsexual may thus present her entire childhood as a somewhat inconsequential "phase" when she was "not really herself."[120] Indeed, as expressed by the conventional Zionist portrayal of the modern Jewish immigration to Israel as a "Return" to an ancient homeland, even eighteen or twenty-five centuries can sometimes be "bracketed off" as a mere interruption of an essentially continuous national project![121]

Ancestry and Descent 3

In addition to the various "bridging" techniques discussed in chapter 2, we also maintain the link between the past and the present through interpersonal contact, with the bridge "connecting" them being embodied by actual people. Such contact is at the heart of fictional encounters between children and celebrated figures from their nation's past,[1] though it is even more effective when it involves real-life encounters with older members of our communities. Furthermore, it is through the "demographic metabolism" allowed by such human bridges that seemingly continuous collective entities such as cities and families are actually *regenerated*.[2] And it is the vision of passing the proverbial torch across those bridges that leads many organizations to use their past members (such as college alumni) to recruit future ones.

As demonstrated by the tremendous public concern displayed throughout President Clinton's 1999 impeachment trial about what the Founding Fathers actually meant when they drafted the U.S. Constitution more than two centuries earlier,[3] our predecessors clearly occupy an extremely important place in our consciousness long after they die.[4] Indeed, as exemplified by their ubiquitous iconic presence on public monuments and paper money (not to mention the Maori tradition of actually enlisting dead ancestors' support before going to war), they often achieve symbolic immortality.[5] Seemingly looking out at them from the colorful murals of their working-class neighborhoods,[6] the ancient Celtic hero Cú Chulainn and the leaders of the 1916 Easter Rebellion have a remarkably "live" presence for the children of Belfast, as did King Solomon, the Hasmoneans, Rashi, and the dozens of other prominent figures from Jewish history who literally surrounded me in the form of the street names of my hometown Tel Aviv.

Dynasties and Pedigree

As one can tell from the way we organize our familial, ethnic, and national identities, our "contact" with past generations is often articulated in biological terms. As we clearly understood long before the rise of modern genetics, the semblance of social continuity is far more compelling when it also involves an element of biological continuity. Indeed, *consanguinity* ("blood") is the functional equivalent of geographical proximity ("place") in the way we mentally construct "natural" connectedness.[7] Our progenitors are thus seen as "prenatal fragments" of ourselves,[8] and some cultures even regard individuals as the personification of all their ancestors.[9]

The role of biology in the social construction of historical continuity is most strikingly evident in essentialist narratives that portray blood "connections" as somehow more real. Presenting their fellow countrymen's blood as the *same* blood that had once flowed in the veins of Ramses II and Akhenaten, early Egyptian nationalists thus stressed the supposedly inevitable "organic" tie bonding them with those ancient pharaohs despite the more than thirty centuries separating them (an argument also made about African Americans by Afrocentrists).[10] The cultural continuity between Egypt's past and present was thus presented as based on the alleged biological continuity between them.

And yet, as reflected in strictly intellectual "dynasties" such as the centuries-long lines of Senegalese Islamic scholars who trace their teachers and teachers' teachers all the way back to the Prophet,[11] not all interpersonal historical connectedness is biological. Indeed, the victory of rabbinical over priestly Judaism and the institutionalization of a pronouncedly celibate priesthood in Catholicism are classic examples of actually preferring such "aristocracy of the mind"[12] to an essentially hereditary aristocracy of the blood. Yet even such "spiritual pedigrees"[13] are ultimately modeled after bloodlines. Given the six-"generation" line of successive sociological mentors stretching from Georg Simmel (through Robert Park, Everett Hughes, Erving Goffman, and myself) to my own students, I indeed envision them as his "great-great-great-grandstudents"!

As the very notion of a "bloodline" seems to imply, the mental linking of past and present generations involves the image of actual "lines" of descent. For instance, the fourteen successive generations constituting the Zildjian "dynasty," which has dominated the world cymbals market since the 1620s,[14] are thus envisioned as collectively forming a single, continuous, linelike mental structure literally called a *lineage*. (Notice again the tremendous significance of names as tokens of sameness. Aside from the common intergenerational repetition of first names that reflects traditional practices of naming newborns after ancestors, sharing a *common surname* helps reify the mental threads that "run through" families across generations, thereby enhancing their perceived continuity.) It was the vision of such a continuous *line of succession*—which implies clear structures of who is "next in line"—that led Spain to restore King Juan Carlos to the throne forty-four years after having deposed his grandfather Alfonso XIII in 1931, and that still inspires die-hard monarchists to keep "lines" of pretenders to the Romanov and even Bonaparte thrones formally alive to this day. Succession is not just based on heredity, however, and is often associated with occupancy of some "office,"[15] as exemplified by the "lines" formed in our minds by series of successive college deans, newspaper editors, or air base commanders.

Note, in this regard, the common image of George W. Bush as the forty-third link in a "chain" of American presidents going back to George Washington. Like other quasi-physical representations of lineages such as family "trees," ropes,[16] and ancestral "rivers,"[17] such a *chain* invokes in our minds the image of a single continuous structure. Yet it also invokes the image of a succession of individuals carrying, as if in some imaginary relay race, the same symbolic baton throughout history. In this way, it clearly underscores the distinctly human phenomenology of *intergenerational transitivity*, which enables us to mentally and experientially transform series of essentially discrete, generationally adjacent pairs (parent-child, teacher-student) into a single continuous "line of succession."

Although they are inevitably strictly diachronic, the relations among the links constituting such *historical contact chains* closely resemble the ties between members of the acquaintance chains we call "small worlds."[18] The sense of interpersonal transitivity that would lead us to hire "someone who knows someone we know" is quite similar to our sense of indirect participation in history "through" ancestors and other historical acquaintances.[19] A young woman who married an 82-year-old Confederate veteran in 1927 was thus regarded by many Southerners seventy-three years later as their "last link to Dixie."[20]

A perfect illustration of such an intriguing sense of intergenerational transitivity is Patricia Polacco's children's book *Pink and Say*, whose protagonist, Say, had once shaken Abraham Lincoln's hand. Polacco ends the book by telling the reader that Say told this story to his daughter, who in turn told it to her daughter, who told it to her son, who told it to his own daughter, who is the author herself, adding that when her father finished telling her the story he showed her his hand, "the hand that has touched the hand . . . that shook the hand of Abraham Lincoln."[21] I have been told that Polacco once read *Pink and Say* to a group of librarians and then shook their hands—and that one of them now invites children to whom she reads the book to shake her hand, which shook the hand that shook the hand that shook the hand that shook the hand that shook the hand that allegedly shook Lincoln's hand!

As we mentally construct such historical contact chains, we often use their length (as measured by the number of "links" constituting a given "chain") as an informal metric for reckoning actual historical distances. In other words, we actually use generations[22] as the chronometrical equivalent of the "degrees of separation"[23] by which we normally measure social distance within "small worlds." Thus, if we think of generations as the twenty-five-year legs of our imaginary historical relay race, we are only twenty "degrees of historical separation" removed from Christopher Columbus.

I am using the word *only* here quite deliberately. Like other mental exercises that involve measuring social distance in terms of degrees of separation,[24] membership in such imaginary relay teams compresses historical distances experientially. As I envision an actual line of twenty people literally linking me to Columbus, he seems somehow closer to me, since "twenty persons away" feels somewhat less distant than "five hundred years ago." I still recall the titillating childhood experience of reading *Memoirs of the House of David*,[25] which portrays Jewish history in such "dynastic" terms, and thinking that I was less than "one hundred and fifty persons away" from Jacob, Moses, David, and other semilegendary biblical figures! It is equally startling to realize that less than forty actual parent-child links take us back to the Norman conquest of England, or that the agricultural revolution started only four hundred generations ago.[26] As the number of mediative links in a historical contact chain decreases, historical distances are experientially compressed. The "small world" thus takes the form of a short history.

Indeed, historical distances seem even shorter once we realize that intergenerational contact need not be confined to adjacent generations. After all, my great-grandmother, who was born in Russia in 1876 and died when I was fifteen, could have actually heard from her own great-grandmother a firsthand account of Napoleon's invasion of Russia in 1812. The thought that I may in fact be only "two conversations away" from a contemporary of Napoleon (as well as of Joseph Haydn, who died in 1809) becomes even more titillating as I realize that when Haydn was born in 1732, Jonathan Swift (1667–1745) was still alive, and that Swift was in turn a contemporary of Thomas Hobbes (1588–1679), who was born when France was still ruled by Catherine de' Medici (1519–89), and that I am therefore actually only seven "degrees of historical separation" removed from Martin Luther (1483–1546), Michelangelo (1475–1564), and Vasco da Gama (1469–1524)!

Yet the length of historical contact chains affects our experience of historical distances in both directions. Increasing the number of inevitably *indirect* contacts we ultimately need to make in order to "access" the past makes such distances seem longer, thereby underscoring the inherent difficulty of trying to convert transitivity to direct contact. After all, the greater the number of intermediate links in a given historical contact chain, the less direct our contact with our ancestors. That also explains why living in a house in which the previous owners lived for only two years makes the presence of *their* predecessors, who actually ate in the very same kitchen and slept in the very same bedroom only three years before us, seem almost "prehistorical."

Yet our experience of historical distances is affected not just by the sheer number of links constituting historical contact chains but also by their actual length, since the longer these links are, the fewer continuity-disrupting "baton passes" they must involve. After all, the fact that I may be historically situated only "two conversations away" from a contemporary of an eighteenth-century composer is not unrelated to the fact that my great-grandmother was eighty-seven years old when she died. The fact that longer "generational" links help to experientially compress historical distances also underscores the tremendous importance of patrilineal forms of organizing descent for social continuity, as they inevitably allow "generations" to exceed women's fertility spans.[27] Furthermore, as exemplified by the fact that during the sixty-eight-year reign of Austrian emperor Francis Joseph I (1848–1916) the White House had no less than seventeen different occupants (Polk, Taylor, Fillmore, Pierce, Buchanan, Lincoln, Johnson,

Grant, Hayes, Garfield, Arthur, Cleveland, Harrison, McKinley, Roosevelt, Taft, and Wilson), such links are equally critical for maintaining organizational continuity.

My experiential proximity to Haydn, not to mention Vasco da Gama, also stems from the fact that the "generational" links constituting historical contact chains are not as discrete as their conventional graphic depiction on genealogical charts might suggest. As my own personal example clearly demonstrates, our life spans often overlap not only with our parents' but also with our grandparents' and even great-grandparents'. Such intergenerational *overlap* also accounts for the mental persistence of social entities such as families and nations, where changes in membership are usually *gradual* and therefore imperceptible.[28] Unlike koalas or woodchucks, humans are not reproductively constrained by a specific breeding season,[29] which means that new members can actually join such groups *continuously*. No "generation" ever replaces another all at once,[30] and different points in the life of a given family or nation often have at least some common members "linking" them. Any changes in the demographic composition of such groups are therefore usually slow, affecting "only a minimum of [their] total life" at any given time and thus remaining proportionally negligible.[31]

As a result, the number of members "connecting" any two moments in a family's or nation's recent history typically exceeds the number of former members who died during that time period as well as the number of new members who have since been born into it. With the exception of extremely unusual catastrophes such as the near-annihilation of the Taino population during the early stages of the European colonization of the Caribbean or the systematic extermination by Nazi Germany of over ninety percent of Poland's three million Jews in the 1940s, such groups regenerate so gradually that we normally experience them as *one and the same* entity that continuously gains some new members while losing some old ones. This perception certainly helps create the illusion of centuries-old families and nations.

Such natural demographic reality is further reproduced by many social organizations (from symphony orchestras to country clubs to professional basketball teams) that make a special effort to maintain *low turnover rates*. Indeed, even in four-year high schools, where dramatic demographic changes affect about one quarter of the student population every year, the other three quarters nevertheless remain intact. The quest for intergenerational continuity also leads some organizations to *stagger* terms in office (as in the United States Senate, where only one-third of the membership comes up for reelection in any given election year)[32] or establish special "transition periods" around major changes of personnel such as top military commanders or presidents. Such continuity is further enhanced by the organizational bylaws, diplomatic treaties, patients' charts, and other documents specifically designed to impersonalize and thereby offset the inevitably disruptive effects of such transitions.[33]

Needless to say, the overall structural imperative underlying all these efforts is *gaplessness*. After all, if one is to envision a continuous royal "line," for example, one should not be able to notice any interregnal gaps between successive monarchs.[34] In fact, any "line of succession" inevitably presupposes such *uninterruptibility* and therefore some temporal overlap, or at least contiguity, between any two successive links.[35] Thus, despite some periods (such as from 304 to 308 and from 638 to 640) when the papal office was actually vacant—not to mention various disruptions of its overall unilinear structure by concurrent "antipopes," such as during "the Great Schism" from 1378 to 1417—we nevertheless envision Pope John Paul II as the 264th link in a supposedly *seamless* apostolic chain extending more than nineteen centuries to Saint Peter. Indeed, it was the Carolingians' ability to effectively present the three centuries from the abdication of the last Roman emperor, Romulus Augustus, in 476 to the coronation of Charlemagne as a "Roman" emperor in 800 as but a temporary "pause" in the life of *the same* continuous political entity that allowed them to lay the genealogical foundations of what would later come to be known as the Holy "Roman" Empire.

As demonstrated by the above—as well as by the equally manipulative practice of "padding" genealogies with phantom generations in order to literally fill gaps in historical chronicles, such as the nine-generation stretch with no corresponding biblical narrative between Noah and Abraham[36]—genealogical chains often seem more seamless than they actually are. Furthermore, achieving such seamlessness may involve strategically glossing over not only genealogical gaps but also various "problematic," continuity-defying links in the chain. During Ehud Barak's July 1999 visit to Washington, for example, in a clear attempt to invoke the image of a continuous string of dovish Israeli leaders, President Clinton thus referred to former prime minister Yitzhak Rabin (1992–95) as Barak's (implicitly immediate) "predecessor," as if trying to tacitly "delete" the frustrating memory of the three-year term of ultra-hawk Benjamin Netanyahu (1996–99), whom Barak had actually succeeded only a few weeks earlier.

Given our unmistakably conservative tendency to glorify the past, we often draw on our ancestors as sources of status and legitimacy seemingly bestowed upon us by the very fact that we "descend" from them, in a top-down manner quite literally reflected in the way it is typically represented in genealogical charts. Emperor Akihito's political legitimacy thus stems from the traditional belief that he is the 125th link in a human chain going back almost twenty-seven centuries to Jimmu, the legendary founder of Japan's royal dynasty, while King Mohammed VI of Morocco's rests on his image as the thirty-sixth great-grandson of the Prophet.[37]

As a "sacred thread linking past and present,"[38] *genealogy* is thus a particularly common system of organizing legitimacy. Enhancing one's social status by establishing descent from revered ancestors (such as claiming to be the seventy-seventh link in a human chain going back more than twenty-five centuries to Confucius)[39] is clearly one of its main functions.[40] Like animal breeders, families use *pedigree* as their key to nobility.[41] Rulers and other members of social elites around the world thus make great efforts to demonstrate their "genealogical worthiness" and even hire special experts to provide them with proper pedigrees.[42] Yet as evidenced by the infamous "one drop rule," whereby having even a single African American ancestor used to be enough in the South to officially be considered black and therefore inferior,[43] such efforts are made only when something is to be gained from the pedigree. Indeed, throughout the Americas, people often cut off entire branches of their essentially multiracial family trees in an effort to fabricate "pure" genealogies that are virtually devoid of any "embarrassing" African ancestors.[44]

Yet demonstrations of "genealogical worthiness" are not confined to biological connectedness and often involve strictly symbolic structures of "ancestry." Throughout his first presidential campaign, for example, Bill Clinton repeatedly invoked Truman and Kennedy in a clear effort to conjure up an image of a chain of popular Democratic presidents implicitly leading to himself. Eight years later, in a meeting with a group of black conservatives, his successor likewise made a point of being joined on the dais by alleged descendants of Thomas Jefferson and his slave Sally Hemings.[45]

Yet pedigree provides more than status: it also provides an identity. Our modern meritocratic ideals notwithstanding, who we are is still also affected by who we descend from, which explains our tremendous obsession with our "roots." "The greatest travesty for African-Americans," it is thus argued, is "that we are disconnected from our beginnings,"[46] and having one's name struck from the genealogical records of one's family was traditionally one of the most dreaded punishments in China.[47] As expressed by the stigma of illegitimacy as well as the terrible identity crisis that often follows the discovery of having been adopted, lacking a clear pedigree is like being "cast out upon [a] sea of kinless oblivion."[48]

Common Descent

Yet the notion of descent "connects" us not only to our ancestors but to numerous contemporaries as well. After all, consanguinity (sharing "the same blood") entails not only lineal ties to parents, grandparents, great-grandparents, and great-great-grandparents but also collateral ties to siblings, cousins, and many other "blood relatives" who also descend from those ancestors.[49]

Common descent is one of the major sources of the commonality on which traditional forms of social solidarity normally rest.[50] Having a common past also entails some general sense of sharing a common present. Rather than envision ourselves as disjointed atoms, knowing that we descend from some *common ancestor* makes us feel somehow "connected."

Such solidarity is the basic sentiment underlying the "communities of blood"[51] commonly known as lineages, cognatic systems, descent groups, or *kinship systems.*[52] As exemplified by the Rothschild "family," the Kennedy "clan," the "House" of Tudor, or the "tribe" of Benjamin, such communities basically consist of everyone who claims descent from some shared ancestor, and the relations among their members ("kin") are therefore ones of *co-descent.* This certainly underscores the indispensability of common ancestors as the social cement holding their descendants together, as is quite evident in family reunions[53] as well as traditional ancestor worship:

> If we think of descent as a tree with the founding ancestor as the trunk . . . we can also visualise the disastrous effect on that tree when the trunk died—the branches would all fall apart as there was nothing left to hold them together. If the tree were to be kept whole, a way of preserving the trunk had to be found: and this in effect is what ancestor worship did, it preserved the founding ancestor without whom there was no connection between the various lines of his descendants.[54]

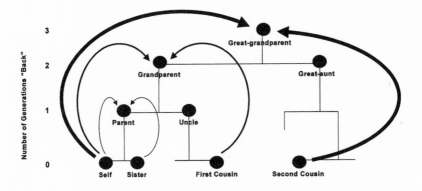

Figure 12 *Genealogical Distance and Ancestral Depth*

History plays a major role in the way we construct kinship, since genealogical "distance," as conventionally measured in degrees of collateral consanguinity, inevitably increases with every reproductive step away from a shared ancestor.[55] We thus regard as "closer" to us relatives whose distance from our shared ancestor is shorter than others'. As we can see in figure 12, genealogical proximity is a function of having a recent common ancestor. Since siblings need to go "back" only one reproductive step in order to identify a common ancestor, they are considered genealogically closer to us than cousins, not to mention second cousins.

The interconnectedness of social and historical "distances" is quite evocatively captured in family trees, those graphic offshoots of the early medieval genealogical diagrams used to determine the possibility of marriage between any two individuals.[56] The roots of these trees signify the common ancestry, and their various branches represent different "levels" of genealogical relatedness. Such interconnectedness becomes even more apparent once we reduce such trees to schematic triangles. As we can see in figure 13, the more recent the historical split between objects, the shorter the social distance between them. *A* is considered "closer" to *B* than to *C* because its historical divergence from the former (at point T_2) was more recent than from the latter (at point T_1). Social distance is thus basically a function of time.[57]

Consider, for example, the use of such distinctly topological reasoning in historical linguistics. Ever since cladograms were first used in the 1850s to depict their genealogical relatedness, we mentally place languages on the

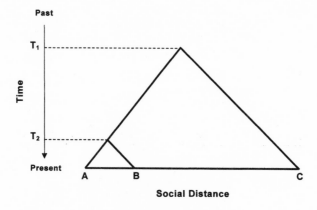

T₁ Point in time when A diverged from C

T₂ Point in time when A diverged from B

Figure 13 *Time and Social Distance*

various "branches" (Baltic, Germanic) of different "families" of languages (Indo-European, Austronesian). The general assumption underlying this unmistakably genealogical imagery is that the more recently languages split off from their common ancestor, the "closer" they are to each other.[58] Thus, while French and Italian are basically regarded as "sister" languages, Czech is considered only a distant "cousin." As we shall see later, such reasoning also pervades the way we nowadays seem to envision our relations with gorillas, zebras, and birds.

Cousinhood, of course, is basically an extension of siblinghood.[59] As such, it includes relations with virtually everyone with whom we share a common ancestor. Indeed, we are "much closer cousins of one another than we normally realize,"[60] and the only question is whether we are first or *forty*-first cousins! As we can see in figure 14, although many of us may have difficulty naming even our third cousins, we are genealogically connected to far more distant "relatives,"[61] despite the obvious fact that our sense of kinship inevitably tends to fade the farther "back" we need to venture into the past to find a common ancestor.[62] Whereas turtles and butterflies do not even recognize grandparenthood, human kinship systems presuppose an unlimited degree of transitivity, so their ancestral "depth" is practically boundless.[63]

As we can see in figure 12, the ancestral depth of kinship systems (as measured in generations separating members from their common ances-

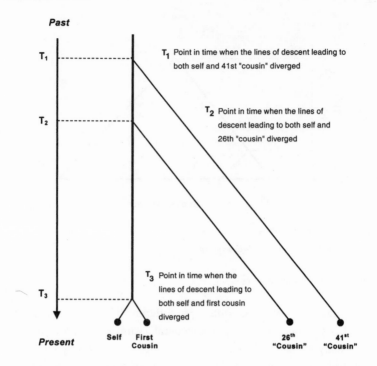

Past

T_1 .... T_1 Point in time when the lines of descent leading to both self and 41st "cousin" diverged

T_2 .... T_2 Point in time when the lines of descent leading to both self and 26th "cousin" diverged

T_3 .... T_3 Point in time when the lines of descent leading to both self and first cousin diverged

Present   **Self**   **First Cousin**   **26th "Cousin"**   **41st "Cousin"**

Figure 14 *Cousinhood and Ancestral Depth*

tor) clearly affects their span (as measured in degrees of recognized cousinhood). The actual size of such systems, in other words, is a function of (and therefore varies proportionally to) their historical depth,[64] progressively expanding every single generational step we go "back." Thus, the larger an extended Chinese family, the more generations of ancestors it evidently worships.[65] In fact, even an entire nation can be viewed as a single extended family,[66] thereby highlighting the critical role of history in the mental construction of social communities ranging from families— through tribes, ethnic groups,[67] nations, and races—to humanity at large. Indeed, the difference between such unmistakably genealogical clusters is only a matter of scale.

The Social Organization of Descent

As we can see in figures 12 and 14, the farther "back" we go in search of common ancestors, the more inclusive our genealogical identity, since a "deeper" sense of kinship inevitably entails a wider range of contemporaries we consider relatives. By invoking Abraham, for instance, as a common ancestor, Jews, Christians, and Muslims can all share a certain sense of symbolic kinship. "We are all children of Abraham," announced New York senatorial hopeful Hillary Clinton, a Methodist, on a 1999 visit to a Jewish cemetery in Morocco,[68] thereby opening a pseudo-genealogical umbrella wide enough to cover herself and her Muslim hosts as well as prospective Jewish voters back home. Such umbrellas should not be too wide, however, as her husband had discovered the previous year upon choosing Uganda as the site of the first-ever public apology for the historical role of the United States in enslaving Africans. "The former slaves are here, not back in Africa," noted one African American New Yorker wryly.[69]

How "deep" we actually go "back" in search of such "common [genealogical] denominators"—as my seven-year-old son characterized his great-grandfather upon meeting his second cousins for the first time—is basically a matter of choice, yet such choices are not just personal. Indeed, it is social conventions that often dictate how many degrees of collateral consanguinity would still qualify people as one's "relatives." Such conventions help curb inordinately endogamous or exogamous interpersonal contact by explicitly disqualifying potential sexual partners who are formally defined as too close to, or too distant from, oneself. Such unmistakably social rules, which basically define our range of kindred recognition, likewise specify whose blood or honor one ought to avenge and who should be invited to weddings and other "family" reunions, as well as the amount of grief one must publicly display when mourning different "levels" of relatives.[70] As one might expect, these rules quite often vary across cultures.[71]

Yet our social environment affects not only the length of the mental threads linking past and present generations but also the very rules according to which we actually "spin" them in our minds. As demonstrated by the unmistakably social conventions and procedures surrounding adoption or surrogate parenthood, the "ancestors" to whom we trace our descent are not always our actual biological progenitors, and it is purely symbolic threads that often connect us to them. Genealogies, in other words, are formal accounts of social rather than strictly natural "descent."[72]

It is thus society, for example, that determines whether we tie ancestry to male or female lines,[73] and nothing could possibly serve as a more striking testimony to the critical role it plays in organizing human descent than the fact that *patrilineality* is the most common path of intergenerational succession worldwide. After all, since biological paternity is always at least somewhat uncertain, a strictly patrilineal organization of descent is inevitably social. In fact, throughout nature, paternity is often quite irrelevant,[74] and many animals do not even know who their fathers are. That would have most probably also been true of humans were it not for the institutionalization of marriage as well as the strict social taboos on female sexual promiscuity, evidently designed to enhance fathers' progenitorial legitimacy. And although some cultures are indeed quite indifferent to paternity and basically choose to adhere to a strictly matrilineal system of organizing descent, almost half of all human societies go to the opposite extreme of officially promoting absolute female-line *genealogical amnesia*.[75] From looking at the biblical genealogies constituting the first eight chapters of 1 Chronicles, for example, one would never guess that women played even a minor role in such multigenerational processes of begetting. Indeed, in strictly patrilineal descent systems women formally have no descendants![76]

Both patrilineality and matrilineality, however, involve only one "line" of ancestors, a perfect manifestation of the unmistakably social nature of the way we organize descent. After all, only society makes us choose between matrilineality and patrilineality as the single genealogical path through which we transmit social rights and duties from one generation to the next rather than draw equally on *both* of them. Only its wish to ensure the continuity of its structure in the least ambiguous (and thus least contentious)[77] way can account for the essentially brutal requirement that we obliterate virtually half of our ancestors from our memory.[78]

Organizing intergenerational succession in a strictly *unilineal* fashion is by no means inevitable, of course. Some societies, in fact, organize descent *ambilineally,* thereby drawing on both male and female ancestors.[79] Whether we actually use systems of single or double descent, however, is an unmistakably social decision.

The "Family of Man"

As I noted earlier, our sense of kinship also extends to humanity at large. Indeed, it now seems quite clear that the proverbial "family of man" is more than just a metaphor.[80] We are, in fact, "a close cousin of practically anybody" else on this planet,[81] to the point where we can actually depict our entire species on a single family tree!

As archaeology and genetics both seem to indicate, we evidently do all descend from a common ancestor, and our differentiation into seemingly separate "races" happened only relatively recently. The genealogical split between Asians and Europeans, for example, seems to have occurred about forty-five thousand years ago, and even the one between Africans and non-Africans is most likely less than a hundred thousand years old.[82] The common ancestors of the present populations of Europe, Asia, Africa, Oceania, and the Americas thus probably lived less than four-thousand generations ago,[83] making Cambodians, Bulgarians, and Mexicans, for example, a "relatively close group of cousins."[84]

Our relatively recent common origins are quite evident from our great genetic proximity. Despite the fact that their population is much smaller and their geographical distribution significantly narrower than ours, genetic differences among chimpanzees or gorillas, for example, are considerably greater than those between Swedes and Nigerians or Samoans and Armenians.[85] Indeed, 99.9 percent of our genes are essentially identical to those of any other human being,[86] and skin-deep differences in pigmentation or eye shape—the result of relatively recent environmental adaptations—are of no biological significance whatsoever.[87] We are thus part of a remarkably homogeneous "genetic fraternity"[88] that includes virtually every human being on this planet:

> Human beings may look dissimilar, but beneath the separate hues of our skins . . . our basic biological constitutions are fairly unvarying. We are all members of a very young species, and our genes betray this secret.[89]

> The progeny of the people who found Australia 50,000 years ago, and the descendants of the tribes who poured down the Americas 12,000 years ago, as well as the heirs to all those other settlers of Europe, Africa, and Asia . . . are all the children of those Africans who emerged from their homeland only a few ticks ago on our evolutionary clock. They may have . . . developed superficial variations, but underneath our species has scarcely differentiated at all. We may look exotic or odd to our neighbors in other countries, but we are all startlingly similar when judged by our genes.[90]

And yet, despite the fact that this pronouncedly *monogenist* vision of all humanity descending from some common ancestor is widely accepted by scientists today, some anthropologists nevertheless advocate an alternative, *polygenist* narrative essentially attributing to the various human "races" altogether separate ancestries. This mnemonic battle over the genealogical identity of the different "races" dates to the Renaissance, when some scholars identified the newly "discovered" natives of the Americas as fellow descendants of Adam and Eve, thereby tracing "racial" divisions to Noah's sons, while others (including Paracelsus, Walter Raleigh, and Giordano Bruno) insisted that they were the descendants of some extrabiblical, protohuman "pre-Adamites."[91] Becoming increasingly popular in the nineteenth century as Europeans' exposure to human diversity began to widen, this polygenist vision of human history was later adapted into the modern language of zoological taxonomy, most pronouncedly manifested in the claim that the various "races" actually constitute distinct species (or even genera)![92]

Racism has always played a major role in polygenism. Henry Fairfield Osborn, a leading proponent of polygenism in the 1920s, was also president of the International Congress of Eugenics.[93] Edward Long, who claimed that blacks and whites actually constitute distinct species, was an antiabolitionist.[94] The separate descent of those two "races" has also been a major theme in the pronouncedly separatist so-called Afrocentrist discourse.[95]

As exemplified by Arthur de Gobineau's claim that "racial" differences were actually fixed "immediately after the creation"[96] (not to mention the very notion of protohuman "pre-Adamites"), one of the distinctive sociomnemonic characteristics of polygenism is the apparent need to push "racial" divisions as far back in time as possible! The greater the antiquity attributed to such divisions, the more compelling one's view of the different "races" as indeed separate from one another.[97] In fact, how far "back" one tries to push the genealogical split among the various "races" is quite indicative of the intensity of one's racist sentiments, since the more recent their common ancestor, the less distant from one another they inevitably seem.

Evolutionary visions of our anthropoid ancestry were thus eagerly embraced by polygenists, as they obviously helped them construct somewhat "safer" genealogical buffers among the various "races." The vision of "parallel series" of humans evolving quite independently from separate species of apes, originally proposed by Karl Vogt as early as 1864,[98] allowed polygenists to link each "race" to an altogether different ape, as quite explicitly manifested in Hermann Klaatsch's and F. G. Crookshank's visions of essentially separate stocks respectively leading from orangutans and gorillas to modern humans of Asian and African descent.[99]

Basically inspired by Franz Weidenreich's idea of the *parallel* evolution of the various "races" from several distinct regional variants of *Homo erectus* and popularized by Carleton Coon's actual portrayal of five seemingly distinct human subspecies that allegedly evolved quite independently from those variants,[100] the modern version of polygenism is indeed often dubbed the "candelabra" model of human evolution.[101] Essentially confining genealogical continuity to different regions, it is thus commonly known as the "multiregional" or "regional continuity" theory.[102] As we can see in figure 15, while monogenists claim that we all descend from protohuman African hominids, multiregionalists believe that *"racial" divisions predate the evolution of Homo sapiens* ("our races are older than our species")[103] and that the different "races" in fact descend from altogether different regional variants of protohuman hominids.

The mnemonic battle over the origin of humans' "racial" divisions certainly affects our notions of kinship. In sharp contrast with what most scientists now believe, the East Asian and Australian lineages envisioned by multiregionalists date to the eastward migrations of *Homo erectus* from Africa more than a million years ago rather than to those of *Homo sapiens* less than a hundred thousand years ago.[104] And whether Australian aborigines indeed descend from bands of *Homo sapiens* who migrated from Africa eighty thousand years ago or from an Indonesian variant of *Homo erectus* who had split off from modern Europeans' ancestral line more than a million years earlier is far from trivial, as it would inevitably determine whether they are indeed my *three*-thousandth or actually only *forty*-thousandth "cousins."

Figure 15 also highlights the fundamental difference between direct ancestors, from whom we can in fact claim descent, and mere *quasi-ancestral*, "dead-end" branches on our family tree, which contain no living descendants, such as all the non-African protohuman hominids in the current monogenist account of our origins. The unmistakably multilinear nature of evolution clearly underscores the ubiquity of extinction:

> Attempts to assert that one or another fossil species is our direct progenitor reflect an outdated notion that evolution is strictly linear and that all fossil forms must be fitted somewhere along a single sequence connecting the past with the

(a)

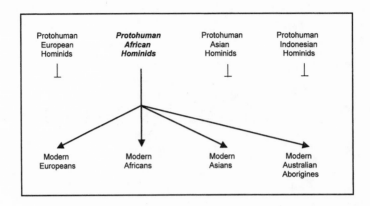

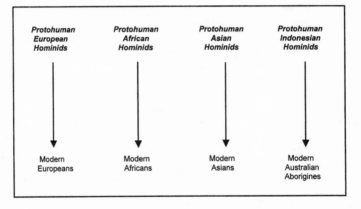

(b)

Figure 15 *Monogenist* (a) *and Polygenist* (b) *Visions of Human Descent*

present. In fact, evolution occurs by a process of repeated branching, with most of the branches becoming extinct fairly rapidly . . . [As such, there is] a large number of parallel evolutionary lines stemming from a common ancestor, only one of which may be represented in the distant future . . . all of the rest having become extinct.[105]

As demonstrated by the mnemonic battle between monogenism and polygenism, the problem, of course, is that it is not always so easy to distinguish our actual great-grandparents from those quasi-ancestral great-uncles and -aunts who did not even leave us any cousins. As Vincent Sarich, an eminent student of our genetic history, once put it, "I know my molecules had ancestors. The paleontologist can only hope that his fossils had descendants"![106]

Given the still-unresolved nature of the genealogical relation between the hominid genera *Homo* and *Australopithecus*, it is thus unclear whether the famous "Lucy," for example, could have actually been my great-grandmother or simply a great-great-aunt. Although Donald Johanson, who discovered her, claimed that her species *(Australopithecus afarensis)* was ancestral to *Homo*, many others believe that it was a mere side branch ultimately leading to a dead end.[107] Similar debates revolve around the precise nature of the genealogical relations between our own species, *Homo sapiens*, and other specific variants of *Homo*. While most anthropologists today regard the East and Southeast Asian variants of *Homo erectus* as great-great-uncles who left no surviving descendants,[108] multiregionalists clearly reject that notion. By the same token, although they were once commonly regarded as our direct ancestors, the so-called Neanderthals are now considered a mere side branch on our family tree that eventually became extinct.[109] Only the apparent need to distance oneself genealogically from allegedly inferior others can account for the Neanderthal ancestry so generously attributed to the Irish by the English and to the Germans by the French.[110]

Apes and Grapes

Yet our sense of kinship need not be confined to hominids. Only some serious *genealogical myopia*, in fact, would make us stop our search for "relatives" at the level of the taxonomic unit conventionally known in biology as the "family."

Following the first reported dissection of a chimpanzee by Edward Tyson in 1699, which already revealed its stunningly closer anatomical resemblance to humans than to monkeys,[111] our biological affinity to apes was formally recognized by science when John Ray, in one of the earliest modern attempts to classify organisms, identified them as *anthropomorpha*, or "man-shaped."[112] This term was soon used by the great taxonomist Carolus Linnaeus to actually place humans and apes within the same zoological order, which he later renamed "primates." In fact, he even placed them within the same genus, featuring the species he named *Homo sapiens* as one of its variants and *Homo troglodytes* (which included the chimpanzee and the orangutan) as another![113] In an essay actually titled "The Cousins of Man," Linnaeus also referred to apes as our "nearest relations."[114]

Such affinity was not explicitly historicized yet, however. The classic vision of all forms of life as the interconnected links of the Great Chain of Being still lacked a clear temporal dimension.[115] And even in the 1750s, when Denis Diderot claimed that species actually evolve historically and Georges Buffon even used a genealogical table to depict the relations among different breeds of dogs,[116] such mutability was still considered strictly intraspecific. The notion of *interspecific transmutation* was implicit in Peter Simon Pallas's attempt in 1766 to portray the affinity between different organisms by drawing a tree, yet this tree still did not include humans.[117]

It was the great naturalist Jean-Baptiste Lamarck who in 1809 first explicitly historicized our affinity to other animals. Not only did he recognize the mutability of species, he also postulated their transmutation into other species: "After a long succession of generations these individuals, originally belonging to one species, become at length transformed into a new species."[118] And although still phrasing it conditionally, he specifically noted our genealogical connection to apes:

> [I]f some race of quadrumanous animals . . . were to lose . . . the habit of climbing trees and grasping the branches with its feet in the same way as with its hands . . . and if the individuals of this race were forced for a series of generations to use their feet only for walking, and to give up using their hands like feet . . . *these quadrumanous animals would at length be transformed into bimanous*. . . . . Furthermore, if the individuals of which I speak were impelled by the desire to command a large and distant view, and hence endeavoured to stand upright, and continually adopted that habit from generation to generation . . . their feet would gradually acquire a shape suitable for supporting them in an erect attitude.[119]

Lamarck's explicitly genealogical vision of our ties to other animals was quite influential, and by the 1850s other European scholars were also postulating our possible descent from apes.[120] Robert Chambers's *Vestiges of the Natural History of Creation*, an 1844 best-seller explicitly describing the process whereby different species "gave birth to one another," ultimately leading to man,[121] was in its fourteenth edition and counted Arthur Schopenhauer, John Stuart Mill, Abraham Lincoln, and Queen Victoria among its readers.[122]

Then in 1859 came Charles Darwin's *Origin of Species,* an entire theory of nature articulated in explicitly genealogical terms. "The Natural System is founded on descent," claimed Darwin, and is therefore "genealogical in its arrangement."[123] The various forms of life share a common ancestor, thereby constituting a "community of descent."[124] This genealogical thrust was also evident in Darwin's extensive use of cladistic imagery to describe biological diversification, basically envisioning the various species branching off from some common ancestor[125] and explicitly portraying their relations as ones between cousins.[126]

Although many of Darwin's readers must have immediately grasped the rather obvious implications of his theory for the evolution of our own species,[127] he himself had not yet made that psychologically as well as theologically provocative deduction explicit in the *Origin.* It was actually Ernst Haeckel who in 1866 finally added that conspicuously absent human angle to the Darwinian argument. Haeckel's explicit discussion of our genealogical relations to all other organisms, along with his evocative use of evolutionary trees to portray our entire biological genealogy, or *phylogeny,*[128] was thus the one that most dramatically propagated Lamarck's original vision of our historical ties to the rest of nature.

In reconstructing our phylogeny, Haeckel also hypothesized an ancient transitional form linking us to apes, thereby introducing "missing links" as a critical element in his distinctly genealogical view of human history. It was, in fact, his vision of such a link, which he named *Pithecanthropus* ("ape-man"),[129] that inspired the actual archaeological search for its remains. And when Eugène Dubois finally unearthed the famous fossil of the so-called Java Man in 1891, he described the species he represented as filling the "void in the series" between humans and apes and even named it *Pithecanthropus erectus* in Haeckel's honor.[130]

As Darwin himself wrote to Haeckel after having read his book, "Your chapters on the . . . genealogy of the animal kingdom strike me as admirable and full of original thought. Your boldness, however, sometimes makes me tremble, but . . . some one must be bold enough to make a beginning in drawing up tables of descent."[131] In fact, he admitted in the introduction to *The Descent of Man* in 1871, he might not have even bothered completing that book had he read Haeckel first![132] This time Darwin was indeed quite explicit about our genealogical ties to other animals, actually claiming that "some ancient member of the anthropomorphous sub-group gave birth to man," who is thus "the co-descendant with other mammals of a common progenitor."[133]

In asserting our genealogical affinity to apes, Lamarck, Haeckel, and Darwin were basically relying on strictly morphological evidence. Their ideas gained further support in the 1960s, however, when Morris Goodman, inspired by earlier findings that human blood closely resembles chimpanzees' and gorillas',[134] set out to compare the molecular structure of our respective blood proteins. Since protein structure reflects genetic structure, it allowed him to measure actual genetic distances and thereby confirm Thomas Huxley's famous claim from 1863 that the African apes are actually closer biologically to humans than to their Asian cousins the orangutan and the gibbon.[135] Even more stunning, however, was his quite unexpected finding that chimpanzees are even closer genetically to humans than to gorillas![136]

In fact, we actually share 98.4 percent of our DNA with chimpanzees.[137] Such remarkable biological affinity is a result (as well as the evidence) of having split off from our common ancestor relatively recently and not having had enough time yet to undergo significant genetic differentiation. After all, "if chimp and human are so alike, we cannot have been evolving separately for very long."[138] As a matter of fact, we can actually date that split by measuring the genetic distance between us. A "molecular clock" devised by Vincent Sarich and Allan Wilson in 1967 helps us calculate the amount of time that has elapsed since the split by comparing the molecular composition of a certain protein that is still present in both of us and consequently estimating the number of genetic mutations that differentiate us from each other.[139] Needless to say, the more similar our molecular makeup, the more recently we must have split off from each other. As we reconstruct our "molecular family tree"[140] based on such calculations, we now envision a common ancestral stem from which the gorilla split off some eight million years ago, most probably followed by the chimpanzee about one or two million years later.[141]

In short, we now estimate that only six or seven million years separate us from apes, six or seven million years since "our DNA and theirs resided in the same cells"[142] and, as one might expect, such relatively short genealogical distance seems to threaten anyone who strongly believes in our distinctiveness vis-à-vis other animals. "One of the ways in which man separates himself from the rest of nature is to put his origins as far back in time as he can,"[143] and the significant compression of the estimated amount of time we now allow for human evolution in our collective memory as a result of Sarich and Wilson's findings is clearly eroding the genealogical buffer that would have helped us (as it certainly helps polygenists) maintain the illusion of such distinctiveness.

Yet our search for relatives need not end even here. While at the taxonomic level of the order my genealogical ties are still confined to other *primates,* such as gorillas and chimpanzees, at the level of the class they also extend to other *mammals,* such as pigs, dolphins, and squirrels. And as I move up to the level of the kingdom it becomes quite apparent that even other *animals,* such as ducks, turtles, and butterflies, are my distant "relatives." Indeed, as we go back in time, our sense of kinship extends well beyond humans and apes to also include rabbits, penguins, frogs, flies, even grapes.[144] As we can see in figure 16,[145] our notion of "cousinhood" should actually encompass every living organism on this planet!

Language and Lineage

Lest we blindly succumb to the seductive spell of biological essentialism, note, however, that even the way we define consanguinity varies cross-culturally as well as historically.[146] To further appreciate culture's considerable impact on how we make sense of nature, note also that, despite the genealogical reality portrayed in figure 16, few of us actually consider grapes or even frogs "relatives."

Rather than simply reflect biological realities, the phylogenetic trees we collectively envision are essentially products of the way we categorize organisms, which is based on unmistakably social traditions and conventions of classifying.[147] As an act of phylogenetic reconstruction, biological classification thus inevitably affects our sociomnemonic vision of human descent.[148]

Consider the way we define humanity. "At what point did our precursors, less and less like ourselves as time recedes, become human?"[149] At what point in history, that is, were mere "hominids" suddenly transformed into full-fledged "humans"? The answer is far from simple. After all,

> [I]t is as legitimate to use the adjective "human" in the inclusive sense . . . as in the exclusive one. . . . Clearly, these two senses of the word are in conflict. . . . Most anthropologists today would lean toward the inclusive use of the term, to embrace the australopithecines as well as later fossil members of the human

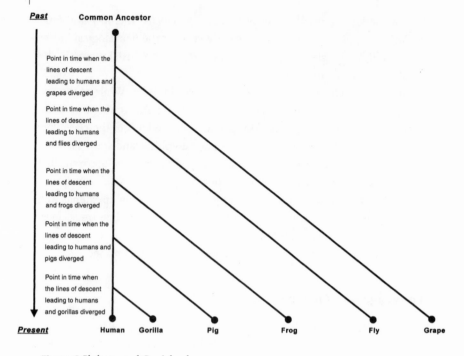

Figure 16 *Phylogeny and Cousinhood*

group: but . . . it might be difficult to view some of [them] as "human" in a functional sense.[150]

Moreover, it is a purely social convention that determines whether our definition is actually based primarily on anatomy (having acquired erect posture, possessing a large brain), behavior (using tools, possessing language), or genetics.[151]

Furthermore, we often may not realize that the actual taxonomic identity of hominid fossils has to do with *language* as much as with biology. Fossils considered by some anthropologists archaic forms of *Homo sapiens* are nonetheless identified by others as advanced forms of *Homo erectus*, and yet by others as an altogether separate species named *Homo heidelbergensis* or *rhodesiensis* depending on whether its remains are found in Europe or Africa.[152] In fact, even the genus *Homo* is just a conventional taxonomic category that "was never more than provisionally defined" in the first place.[153] Since there is, after all, "no non-arbitrary yardstick available for the genus as reproductive isolation is for the species,"[154] genera are essentially products of unmistakably conventional taxonomies. Thus, it may have very well been our general cultural bias toward tool making—not to mention "generic pride"—that led Louis Leakey to identify the producer of the earliest stone tools as *Homo* (*habilis*) rather than *Australopithecus*,[155] and some of his critics in fact claimed that it was just an advanced form of the latter.[156] Indeed, when his son Richard later identified another fossil as *Homo habilis,* even his own coauthor insisted that it was an australopithecine![157]

Unmistakably social conventions of classification likewise affect the number of genera that, along with *Homo,* constitute the taxonomic family "hominids." It is still unclear, for example, whether "gracile" and "robust" australopithecines are indeed just two subgeneric forms of *Australopithecus* or perhaps two entirely distinct genera, *Australopithecus* and *Paranthropus.*[158] Although the fossil originally identified by Robert Broom in the 1930s as *Australopithecus transvaalensis* was very soon renamed by him *Paranthropus robustus,*[159] some anthropologists still reject Broom's later generic designation and essentially consider *Paranthropus* a merely subgeneric ("robust") form of *Australopithecus.*

Indeed, even the designation of the taxonomic level at which we actually distinguish humans from apes is anything but definitive. The fact that we are formally assigned to three separate families (Hominidae, Pongidae, and Hylobatidae) is a purely conventional taxonomic arrangement that probably warrants some serious reconsideration given our remarkably close genealogical tie to gorillas and chimpanzees.[160] Since it is only a "narrow genetic crevice" (rather than a wide "biological abyss") that separates us from them,[161] there is no reason for us, for example, not to consider ourselves apes.[162] Nor, for that matter, is there any biologically compelling reason for chimpanzees not to be included in the hominid family.[163] Jared Diamond, in fact, assigns us even to the same *genus,* essentially claiming that we are "just a third species of chimpanzee"![164] Echoing Darwin's contention that "if man had not been his own classifier, he would never have thought of founding a separate order for his own reception,"[165] he thus reminds us that only our anthropocentricity prevents us from realizing that the infinitesimal (1.6 percent) genetic distance separating humans from chimpanzees is even shorter than the distance between different species of gibbons.[166]

Taxonomic disputes typically underscore the inevitable tension between two contrasting principles of classifying, namely lumping and splitting.[167]

Stressing similarity, "lumpers" promote "taxonomic deflation"[168] by essentially downplaying differences, as so clearly manifested in their efforts to reduce *Paranthropus* to a mere subgeneric form of *Australopithecus*, portray the Neanderthals as just a regional variant of *Homo sapiens*, and even challenge the conventional distinction between *Homo* and *Australopithecus*.[169] "Splitters," by contrast, play up the differences between the Neanderthals and modern humans and regard the hominid inhabitants of Europe 400,000 years ago as a distinct species *(Homo heidelbergensis)* rather than as some archaic form of *Homo sapiens*.[170] They likewise consider the taxonomic categories *Australopithecus* and *Homo* "highly inflated," basically contending that the significant morphological variance within each of them clearly warrants a formal introduction of additional hominid genera.[171]

To help them add a certain aura of inevitability to the mental divides separating one "kind" of relatives from another, splitters often assign each of those clusters a separate name. Language is a very effective splitting device,[172] and assigning different sets of objects different names makes them seem more distinct from one another. Using taxonomic labels such as *Homo mediterranaeus* and *Homo europeus*, for example,[173] certainly made it easier for polygenists to portray the various human "races" as separate species.

Yet language can also help promote affinity in that referring to chimpanzees as our "cousins," "closest relatives," or "sibling species"[174] makes them seem somehow closer to us, and assigning things a common label generally makes them seem more similar to one another.[175] Identifying the Neanderthals as *Homo sapiens neanderthalensis* instead of *Homo neanderthalensis*, for example,[176] thus helps lumpers transform them from distant generic relatives into essentially conspecific cousins! Indeed, it is the titillating prospect of virtually obliterating the utterly conventional mental gulf traditionally separating humans from other animals that leads adventurous authors to give their books such provocative titles as *The Naked Ape* or *The Third Chimpanzee*.[177]

Drawing on Greek *(anthropos, pithekos)* and Latin *(homo, simia)*, lexical compounds likewise help promote the mental fusion of "human" and "ape." Thus, when Raymond Dart introduced in 1925 the first australopithecine fossil, he actually assigned it to a new family of primates which he formally named *Homo-simiadae*,[178] thereby essentially employing the same taxonomic strategy used by Haeckel when naming his hypothetical "missing link" *Pithecanthropus*.[179] Needless to say, this highly evocative lexical hybrid was clearly designed to help him explicitly conjure some "intermediate between living anthropoids and man," a "man-like ape."[180]

As Darwin himself noted,[181] both lumping and splitting presuppose a choice between noticing similarities or differences, *both* of which can in fact be found between any two things we compare.[182] The choice is thus often a matter of social convention—another useful reminder that the way in which we actually "map" nature and its history is ultimately social.

Historical Discontinuity 4

We have thus far examined some of the ways in which we try to generate the mnemonic experience of historical continuity. Yet such attempts are quite commonly offset by diametrically opposite efforts to create the experience of historical *dis*continuity. And whereas the kind of mnemonic "editing" presupposed by the former is geared to deliberately overlook actual temporal gaps between noncontiguous points in history, the one involved in the latter is specifically designed to help transform actual historical continua into series of seemingly unattached, freestanding blocks of time. As we can see from contrasting figures 17 and 11, instead of mnemonic "pasting," historical discontinuity thus involves some mnemonic "cutting," since rather than try to project a semblance of gaplessness, the goal is to promote a vision of actual historical gaps.[1]

Instead of envisioning history as an uninterrupted chain of essentially contiguous occurrences flowing into one another like the successive musical notes that form legato phrases, we are now dealing with a mnemonic vision featuring actual ruptures between one chunk of history and the next, resembling the musical pauses between the successive notes that form staccato phrases. The contrast between these two diametrically opposite sociomnemonic visions of the past is quite evident in paleontology and geology, where gradualist narratives featuring graded chains of intermediate organic forms evolving from one another almost imperceptibly are contrasted with *episodic* ones featuring supposedly *discrete* historical "eras" ("epochs," "ages") separated from one another by pronouncedly sharp *breaks*.

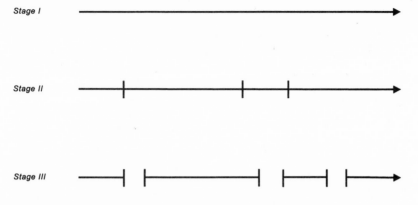

Figure 17 *Mnemonic Cutting*

As we have seen in chapters 2 and 3, legato narratives are essential for projecting a sense of historical continuity. As we shall now see, staccato narratives are equally indispensable to any efforts to generate a sense of historical *dis*continuity.

The Social Punctuation of the Past

As one might expect, constructing such a pronouncedly discontinuous vision of the past involves producing the mnemonic equivalent of orthographic spacing or musical phrasing. In order to fully understand this process, we must therefore identify the structural and functional mnemonic equivalents of commas, spaces between words, pauses, and rests. Such punctuation devices are at the heart of the sociomnemonic process commonly known as *periodization*.

The purportedly distinct "periods" explicitly articulated through this process are typically delineated by historical events collectively remembered as major *watersheds* in the lives of specific mnemonic communities. As graduating from college or getting married is for individuals, such events help carve out significant "chapters"[2] in the lives of those communities by essentially marking when they begin and end. Thus, for many Hutu during the 1980s, the 1972 killing of tens of thousands of their people in Burundi was a cataclysmic event that practically separated the "premassacre years" from everything that happened since.[3] In a somewhat similar vein, for many Britons the death of Queen Victoria in 1901 marks the dawn of the modern age.[4] In fact, many nations formally incorporate such events into their collective memories by designing special holidays to commemorate them. The French evacuation of the Bizerte naval base on 15 October 1963, which is annually commemorated on Tunisia's Evacuation Day, and the nationalization of Iran's oil industry on 20 March 1951, which it commemorates annually on Oil Nationalization Day, are classic examples of such historic "turning points."

Some of these "historic" moments, however, come to be defined as significant watersheds only retrospectively. Events we now regard as marking "defining moments" may not have even attracted much public attention when they actually occurred. For example, when one considers the shootout between South African troops and some South West African rebels at Omgulumbashe on 26 August 1966, or what began as a simple student demonstration against the establishment of Urdu as the official language of the predominantly Bengali-speaking province of East Pakistan on 21 February 1952, one realizes that only in retrospect did the events presently commemorated on Namibia's Heroes' Day and Bangladesh's National Mourning Day come to be seen as such pivotal watersheds. By the same token, only with historical hindsight can one recast a failed guerrilla attack which took place on 26 July 1953 as the beginning of what would come to be known years later as the Cuban Revolution![5]

Such events are generally regarded as "benchmark episodes"[6] that mark the transition from one supposedly distinct chapter in a mnemonic community's history to the next because, like the days on which we got our driver's license or lost our virginity, they are collectively perceived as having involved significant identity transformations.[7] The official change of Dahomey's colonial name to Benin on 30 November 1975, which is annually commemorated on National Day, is a classic example of such a transformative event. So are the proclamation of Poland's first constitution on 3 May 1791 and the overthrow of the monarchy in Libya on 1 September 1969, which are annually commemorated on Constitution Day and Revolution Day, respectively.

A major event that is often collectively remembered as a significant historical watershed is a nation's political "birth" following a merger of several smaller units (as in Switzerland in 1291 or the United Arab Emirates in 1971) or, as is more often the case, a national struggle for independence. Indeed, of the 191 countries whose national calendars I have examined, 139 celebrate a national "birthday" commemorating the historic moment when they became formally independent, and some (Algeria, Uruguay, Mozambique, Eritrea) also commemorate the day on which their national struggle for liberation was actually launched. Six of Angola's seven national holidays specifically designated to commemorate major historical events (Armed Struggle Day, Pioneers' Day, Armed Forces Day, Independence Day, Victory Day, and Heroes' Day), in fact, actually revolve around its struggle for independence from Portugal between 1961 and 1975. Multiple annual commemorations of the "births" of Panama (six), Ecuador (five), and Haiti (five) likewise underscore the significant role of nations' political birth as a historical watershed.

It is specifically as a form of classification that periodizing helps articulate distinct identities, and the way men and women respectively use career moves and births of children as autobiographical benchmarks, for example,[8] certainly underscores the fundamentally different manner in which they normally organize their identities. Temporal discontinuity is a form of mental discontinuity,[9] and the way we cut up the past is thus a manifestation of the way we cut up mental space in general. In the same way that "holy days" help concretize the moral distinction between the sacred and the profane[10] and weekends help give substance to the cultural contrast between the public and private domains, the temporal breaks we envision between supposedly distinct historical "periods" help articulate mental discontinuities between supposedly distinct cultural, political, and moral identities. The conventional Zionist distinction between the Jews who lived in Palestine before 1882 ("the old *yishuv*") and those who emigrated there since then ("the new *yishuv*") is thus clearly more than just chronological, as it actually helps articulate the cultural and political contrast between the traditional-religious and secular-national worlds.[11] And in the same way that the Exodus marks the fundamental moral discontinuity between idolatry and monotheism,[12] the temporal break we envision between "pre-Columbian" and post-1492 America helps flesh out the major cultural contrast between "indigenous" and "European."[13]

Assimilation and Differentiation

As we classify things, thereby arranging them in seemingly distinct mental clusters, we normally allow the perceived similarity among the various elements constituting each cluster to outweigh any differences between them. As a result, we come to regard those elements as somewhat interchangeable variants of an essentially homogeneous mental entity. At the same time, in order to enhance our perception of different clusters as distinct from each other, we also tend to inflate the perceived mental distance between them.[14]

Like any other form of classification, periodizing thus presupposes a pronouncedly nonmetrical, *topological* approach[15] that highlights relations between entities while basically ignoring their internal makeup. That entails a somewhat plastic experience of temporal distances that involves mnemonically compressing those *within* any given conventional "period" while inflating those *between* periods. As we can see, although utterly irrelevant metrically, the difference between *intra-* and *inter-* is critical when approaching reality topologically.

The first of these twin mnemonic processes, *historical assimilation*, involves assigning each of those conventional blocks of history a single common label such as "Neolithic" (farming), "eighteenth-century" (literature), or "Ming" (art). As a result of the sociomnemonic habit of downplaying intraperiodic variance to the point where we regard each such "period" as practically homogeneous, we also attribute to it a single, essentially uniform identity. We may thus come to conventionally associate the entire 1,800-year "Exile" chapter in Jewish history with persecution[16] and collectively remember more than five centuries of European history as "dark."

As we can see in figure 18, such schematic visions of history also result in a mnemonic compression of temporal distances within any conventional "period." We may thus come to perceive "Renaissance" artists like Donatello (whose early work dates to the 1410s) and Titian (who was still painting in the 1560s) as contemporaries, and forget that "medieval" luminaries such as Saint Benedict (480–547) and Chaucer (1340–1400) actually lived more than eight centuries apart from each other. Along similar lines, as we crudely identify anything that existed in the Western Hemisphere before the arrival of Europeans as "pre-Columbian,"[17] we tend to conflate the Olmec and Aztec civilizations of Mesoamerica (or their Chavín and Inca counterparts in the Andes), which actually flourished two thousand years apart from each other, often forgetting that they were in fact as historically remote from each other as present-day Italians are from the ancient Romans. Lumping together nearly three thousand years of pre-Ptolemaic northeast African history in the single unit conventionally remembered as "ancient Egypt" likewise implies forgetting that, like Chaucer and the Aztecs, the last pharaohs of the Thirtieth Dynasty were actually a couple

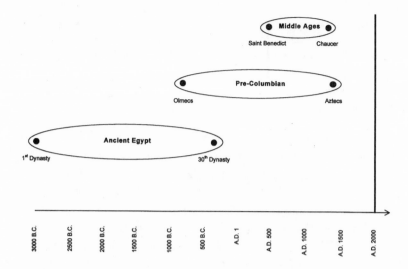

Figure 18 Historical Assimilation

of centuries closer to *us* than they were to those who founded Egypt's First Dynasty!

Yet like any other form of classification, periodizing the past involves not just *intraperiodic lumping* but also *interperiodic splitting*. Not only do we attribute to an entire historical "period" a single, uniform identity, we also attribute separate identities to what we consider "separate" periods, as exemplified by the Zionist portrayal of the "Exile" period in Jewish history as essentially antithetical to the periods immediately preceding and following it.[18] Historical assimilation is thus typically complemented by the diametrically opposite sociomnemonic process of *historical differentiation*.

"Periodizing" the past basically involves a mnemonic transformation of actual historical continua into seemingly discrete mental chunks such as "the Renaissance" or "the Enlightenment." Indeed, it is our ability to envision the historical equivalents of the blank spaces we conventionally leave between the different chapters of a book or at the beginning of a new paragraph[19] that enhances the perceived separateness of such "periods" and gives the aforementioned watershed metaphor such resonance. Like the actual river separating March from April in Saul Steinberg's cartoons,[20] it is the imaginary stretches of historical void separating 1979 from 1980,

Figure 19 *A Topological History of the Western Hemisphere*

for example, that promote our memory of "the '70s" and "the '80s" as such distinct historical entities. As we can see in figure 19, it is our mental vision of the quasi-geological fault separating 1491 from 1493 that likewise helps us remember the "pre-Columbian" and "American" chapters of the history of the Western Hemisphere as distinct "eras."[21]

Such perceived gaps clearly affect the temporal distances we envision separating different historical "periods" from one another. In order to promote the sociomnemonic vision of two contiguous yet conventionally "different" chunks of history as actually discrete, we tend to inflate the imaginary divides supposedly separating them from each other.[22] As a result, crossing such "historical Rubicons" transforms metrically small steps in physical time into topologically giant leaps[23] in social time—in the same way that we are instantaneously transformed from "minors" into "adults" when we turn eighteen and that someone who has had only one sexual experience is perceived as somehow "closer" to one who has had thirty-seven such experiences than to a virgin.[24] In order to help maintain the illusion of wide historical gaps actually separating "different" periods from one another, we thus mnemonically inflate the distance between everything that happened prior to the particular "watersheds" marking their boundaries and everything that has happened since. As a result, we come to perceive the distance from 1491 to 1493 as considerably longer than the metrically identical distance separating 1491 from 1489. After all, as we can see in figure 19, whereas 1489 and 1491 are both part of one "era," 1491 and 1493 are conventionally perceived as straddling a wide historical Rubicon separating that "era" from the following one.[25]

As if to actually reify the distance between them, we also literally place different historical "periods" in different chapters (or even in different sections) of history textbooks as well as separate wings of museums, thereby helping give substance to the imaginary divides separating them from one another. Such spatial segregation certainly helps perceive those purely conventional figments of our mind as distinct, altogether separate "eras."

History and Prehistory

The attribution of separate identities to contiguous yet conventionally separate historical "periods" is often manifested in the way we perceive them as being in opposition to each other, as exemplified by the way many Americans seem to view the past and present chunks of history that respectively ended and began on 11 September 2001. Nowhere are such perceived contrasts more profound than when we quite self-consciously try to establish what we hope will come to be remembered as the *beginning* of a new "era," a highly ambitious sociomnemonic act that epitomizes the process of historical periodization.

Explicitly playing up (often to the point of exaggeration) the perceived contrast between two contiguous yet conventionally separate historical "periods," establishing a new beginning usually presupposes the death of some prior entity. When George W. Bush announced at the 2000 Republican Convention that "it is a time for new beginnings,"[26] he also was implying the anticipated death of the Clinton-Gore "era." By the same token, when President Gamal Abdul Nasser told fellow Egyptians in 1956 that on the day after the British evacuation of the Suez Canal they would awaken to "a bright new era,"[27] he was also announcing the impending death of imperialism.

Given all this, establishing a "new beginning" often involves destroying every possible link to anything that preceded it. Indeed, as exemplified by the actual scope of the French and Russian Revolutions, social revolutionaries often try to virtually obliterate the existing social order before proceeding to establish a new one in its place. It was his apparent wish to dramatize the imaginary gap separating the young Turkish society from its recent (and therefore potentially still dangerously "contagious") Ottoman past that led Atatürk in the 1920s to move the official seat of government from Istanbul to Ankara, formally abolish the Mohammedan calendar and traditional Arabic script, outlaw the use of the fez and the veil, and practically purge the Turkish language of any Persian influence.[28]

Consider also the ritual haircut that marks the transition from civilian to military life, or the formal renaming of religious converts, slaves, and nuns. Such *rites of separation*[29] are specifically designed to dramatize the symbolic *transformations of identity* involved in establishing new beginnings, essentially implying that it is indeed quite possible to "turn over a new leaf" and be somehow "reborn." This possibility is often manifested in explicit allusions to "revitalization" and rejuvenation,[30] not to mention actual efforts to socially engineer a new type of person who would embody the dramatic historical break between the old and new "eras," as in the highly ambitious Zionist attempt to replace the old "exilic" Jew by the young Israeli *sabra*.[31]

In order to effectively project a sense of historical discontinuity, one also needs to destroy the mental "bridges" we discussed earlier. Indeed, establishing new beginnings involves various sociomnemonic practices that are the exact opposite of the ones we use to promote a sense of continuity. It was the special mnemonic significance of place, for example, that led the Assyrians to systematically uproot vanquished populations from their lands, and the evocative role of ruins that led the Spaniards to virtually raze the Aztec city of Tenochtitlán—as the Romans did to Carthage after the Third Punic War—before proceeding to build Mexico City on the very same site.

By the same token, it is the mnemonic significance of relics and anniversaries that leads victorious armies and new regimes to destroy historical monuments and remove certain holidays from the calendar. Thus, Hungarians no longer commemorate their liberation by the Soviet army in 1945, and South Africa no longer feels compelled to pay annual homage to Paul Kruger.[32] It was unmistakably sociomnemonic considerations such as these that also made Romanians tear the socialist emblem out of their national flag in 1989. They have likewise led new regimes to quite self-consciously change their countries' national anthems[33] and rename streets[34] and cities (such as from "Petrograd" to "Leningrad" and back to "Saint Petersburg"), as well as entire countries (such as from "British Honduras" to "Belize").

In marking significant historical breaks, "watersheds" often serve as extremely effective *chronological anchors*, which is why throughout 1980 some U.S. television networks would pointedly end their evening news by counting the number of days that had passed since the takeover of the American embassy in Tehran.[35] As Mark Twain described the sociomnemonic role of the American Civil War in the South,

> [T]he war is what A.D. is elsewhere: *they date from it*. All day long you hear things "placed" as having happened since the waw; or du'in' the waw; or befo' the waw; or right aftah the waw; or 'bout two yeahs or five yeahs or ten yeahs befo' the waw or aftah the waw.[36]

Even more spectacular in this regard is the sociomnemonic role of the birth of Jesus circa 4 BC and the flight of Mohammed from Mecca to Medina (the *hegira*) in AD 622 as the "pivotal" foundations of conventional chronological dating frameworks,[37] the "hinge[s] on which the door of history swings."[38] As quite effectively illustrated by the dramatic break we seem to envision between the periods we respectively designate by the letters "BC" and "AD," it is as if history indeed began on the first year of our standard chronological era![39]

The common image of such events as *historical points of departure* is quite evident from their association with the exceptionally grandiose sociomnemonic practice of explicitly *resetting a mnemonic community's "historical chronometer" at zero*.[40] Consider, for example, Cambodian dictator Pol Pot's megalomanic decision to designate 1975, the year in which he came to power, as "Year Zero," or the concept of "Zero Hour" *(Stunde Null)* used by some Germans in 1945 in an effort to project an altogether new political identity based on a clean break with their nation's irrevocably tainted recent Nazi past.[41] Along similar lines, 1916, the year of the Easter uprising against Britain, is sometimes viewed as "the year one in Irish history."[42] Even more spectacular was the attempt made by France in the 1790s to formally replace the conventional Christian Era with a pronouncedly French "Republican Era" that began with the foundation of the First French Republic on 22 September 1792,[43] a remarkable sociomnemonic experiment repeated in the 1920s by the Fascists, who likewise introduced throughout Italy a new standard chronological era that began with their historic March on Rome in October 1922.[44]

Resetting "historical chronometers" at zero typically also involves emphasizing *primacy,* as when the first weekday following the historic 1979 referendum affirming the foundation of Iran's Islamic Republic was explicitly proclaimed by the country's supreme leader, Ayatollah Ruholla Khomeini, "the *first* day of a government of God."[45] Consider also in this regard the conventional Zionist depiction of the eastern European Jews who came to Palestine in 1882 as Israel's "first" immigrants *(ha-aliyah ha-rishonah),* further reinforced by their standard portrayal as the country's "Founding Fathers" or "Pioneers." Indeed, they themselves were quite self-conscious about their future historical image, even naming two of their first settlements Rishon Le-tziyyon ("first to Zion") and Rosh Pinnah ("cornerstone").

Needless to say, remembering the Jews who came to Palestine in 1882 as that country's "first" settlers also implies a *mnemonic obliteration* of every

Jew who had ever emigrated there before that—not to mention all those who had never even left the country during the eighteen centuries conventionally portrayed in Zionist historiography as the "Exile" period, when all Jews were supposedly living outside their homeland.[46] Furthermore, this depiction implicitly entails suppressing the memory of all the non-Jews who were living there when those Jewish immigrants arrived, thereby helping project an unmistakably Eurocentric view of pre-1882 Palestine as a virtually empty, desolate[47] place waiting to be settled by those "pioneers."

Indeed, such *mnemonic myopia*[48] is quite common in colonial discourse involving "settlement." Thus, despite the fact that the ancient sagas explicitly noted that when the first Norwegians arrived in Iceland in the ninth century they found that Irish monks had already preceded them, their commitment to the island's pronouncedly Scandinavian identity nevertheless led them to essentially disregard such pre-Scandinavian Celtic presence and thereby present those Norsemen as its "first" settlers![49] By the same token, despite the fact that the first British settlement in Australia was established at least forty thousand years after the island had already been "aboriginally" settled, the national Australian holiday commemorating its establishment in 1788 is nevertheless called Foundation Day.

Mnemonic obliteration of entire populations is also quite common in discovery narratives. When the *New York Times* offers its readers a brief historical profile of Mozambique that begins in 1500 with the arrival of the Portuguese, it implicitly portrays that country as virtually empty at the time of its "discovery," thereby essentially relegating its entire pre-European past to official oblivion.[50] And when we say that Columbus "discovered" America, we are basically implying that no one was there before him, thus implicitly suppressing the memory of the millions of Native Americans who were actually living there at the time of his arrival.

As demonstrated by the way we conventionally label anything that existed in America prior to Columbus's arrival as "pre-Columbian,"[51] 1492 marks a fundamental break between America's actual "history" and what we apparently consider its mere *prehistory*. (Along somewhat similar lines, in the unmistakably Christocentric folk historiography of Ireland, anything predating Saint Patrick's celebrated arrival circa 432 is basically dismissed as "pagan prehistory.")[52] As provocatively implied by the title of Noam Chomsky's scathing 1993 critique of European imperialism, *Year 501*,[53] the cultural entity we call "America" is commonly perceived as having been "born" on 12 October 1492. Anything that happened throughout the Western Hemisphere prior to that date can therefore only be part of some "pre-America."

Essentially regarded as a mere prologue to its actual history, much of America's "prehistory" is thus *forgotten*. Consequently, the Norse voyages to Greenland, Newfoundland, and possibly also Labrador and Nova Scotia in the late tenth and early eleventh centuries are not considered part of the standard narrative of its "discovery."[54] Although most of us are quite aware of those early crossings of the Atlantic five centuries before Columbus, we still regard his celebrated landfall in the Bahamas as the formal beginning of America's history. And if America was indeed only "born" on 12 October 1492, nothing that had happened there prior to that date can actually be considered part of "American history."

As exemplified by the traditional image of the creation of the world *ex nihilo*, we tend to envision beginnings as preceded by actual void. Thus, in order to dramatize the historical break between Jews' former life in exile and "new beginnings" in their homeland, native-born *sabras* were sometimes portrayed in early Israeli literature as orphans.[55] For the same reason, practically disregarding their early years in "exile," the Zionist narrative often also presented immigrants' lives as starting only upon their arrival in Palestine![56] The very existence of such *prehistorical void* helps remind us that establishing historical "beginnings" always presupposes an element of *amnesia*. Thus, as Americans come to remember the colonization of New England in 1620 as the beginning of the European settlement of the United States, for example, they implicitly also come to forget the colonization of Virginia in 1607, not to mention the Spanish colonization of Florida in the 1560s and New Mexico in the 1590s.[57] As one of Zionism's leading visionaries put it quite bluntly,

> [W]e cultivate oblivion and are proud of our short memory. . . . And the depth of our insurrection we measure by our talent to forget. . . . The more rootless we see ourselves, the more we believe that we are more free, more sublime. . . . It is roots that delay our upward growth.[58]

As we can see in figure 20, establishing any beginning presupposes an implicit agreement to disregard anything that predates it as somehow "irrelevant" and therefore immemorable. Such seemingly innocuous yet unmistakably brutal *mnemonic decapitation*[59] is designed to help promote

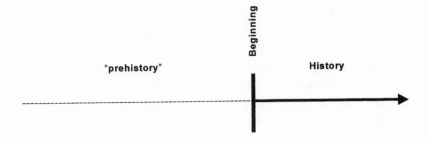

Figure 20 *Mnemonic Decapitation*

the fundamental discontinuity between what we consider history and what we regard as prehistory and thereby tend to forget because it is conventionally considered irrelevant.[60]

The expectation that we can actually disregard anything predating a certain point in time is evident in laws that absolve businesses from any debts they may have had prior to declaring bankruptcy. It is also evident in statutes of limitations, the ultimate manifestation of the notion that it is indeed possible to formally put certain parts of the past "behind us."[61]

When trying to put things "behind us," we do not necessarily deny that whatever predates a certain historical point of departure actually happened. Yet by establishing certain "phenomenological brackets" we somehow relegate these events to social irrelevance. The tacit distinction between history and "prehistory" implies that, like a preface of a book or "introductory" remarks in lectures, it basically lies outside the official historical narrative and, as such, is normatively excluded from what we are expected to remember. Statutes of limitations imply that even what we all agree actually happened can nevertheless be formally banished to some "prehistorical" past that for all practical purposes is considered irrelevant and therefore officially forgotten. Thus, as students transfer from one college to another, for example, they can in fact expect their new school to formally "forget" their old grades when calculating their cumulative grade point average.

Consider also the quest of many Germans today for "normalcy unburdened by history,"[62] or Montenegrin president Milo Djukanovic's recent appeal to Croats to put Yugoslavia's 1991 war against Croatia "behind them."[63] Such eagerness to basically "clip" the past so as to quite conveniently be able to start mnemonically "afresh" was also displayed by former Khmer Rouge leader Khieu Samphan, who in 1998 asked Cambodians to essentially "forget the past" and "let bygones be bygones."[64] (His effort to practically wipe out the memory of the massacre of more than one million Cambodians only twenty years earlier was officially endorsed by Prime Minister Hun Sen, who urged his countrymen to "dig a hole and bury the past and look ahead to the twenty-first century with a *clean slate.*")[65] Similar calls to "turn over a new leaf" and put certain things "to rest" by essentially obliterating their memory are also made by advocates of ex-convicts' right to start "a new life" untainted by their criminal past. Applying the sociomnemonic logic underlying the norms of declaring bankruptcy, many of those who in 1998 opposed the execution of Karla Faye Tucker in fact claimed that as a born-again Christian she should not be held accountable for a murder she had committed prior to her spiritual "rebirth"!

The Social Construction of Historical Discontinuity

Yet as we are occasionally reminded by poems and books that begin, quite provocatively, in the middle of a sentence,[66] historical discontinuity should in no way be regarded as a given. Like cropping photographs, carving conventional "periods" out of their historical surroundings is an artificial act and, as such, far from inevitable. Thus, although most Israelis, for example, consider the foundation of their state in 1948 a virtually indisputable "watershed" (indeed, a popular account of the events of that year is even subtitled *Between the Eras*),[67] it is actually a nonevent for Israel's largely apolitical ultraorthodox community. (Challenging the Zionist wish to detach modern Israeli history from its immediate past, some Israeli historians likewise question the conventional distinction between the "old" pre-1882 and "new" Jewish communities in Palestine.)[68] By the same token—as we are quite effectively reminded by the sarcastic remark that "one wonders how the Nez Percé and Navajos survived the boredom of long centuries waiting for invaders from the East to show up"[69]—nor, for that matter, is 1492 actually perceived as a historical point of "departure" by Native Americans, whose ancestors had been living in America for thousands of years before it was finally "discovered" by Europe.[70]

Indeed, the perceived reality of the seemingly discrete segments into which we conventionally carve the past is a product of the historical gaps we collectively envision separating them from one another. Yet such cleavages, so obvious to anybody who has been mnemonically socialized into a particular tradition of "periodizing" the past, are virtually invisible to anyone else! After all, in the real world, there are no actual gaps separating the impressionist and cubist "periods" in Western art (which actually overlapped with each other)[71] or France's "Fourth Republic" and "Fifth Republic" (which ended and began, respectively, on the same day) from one another.[72] Cutting up the past into supposedly discrete "periods" is basically a mental act and, as we shall now see, it is usually done with an unmistakably social scalpel.[73]

Much of the construction of historical discontinuity is, in fact, tacitly accomplished through language. Whereas attaching a single label ("medieval") to more than ten centuries of European history helps us perceive them as a relatively homogeneous block of time, assigning each conventional "period" a different label helps us split them apart in our mind as different and therefore also *separate* chunks of history. In the same way that it helps us mentally separate "childhood" from "adolescence" and "winter" from "spring,"[74] language thus also enhances our vision of actual historical gaps separating "Mesolithic" from "Neolithic" tools and "Renaissance" from "baroque" music. In a similar vein, distinguishing "archaic folk" from "early moderns" highlights the historical divide supposedly separating East Asia's Lower and Middle Paleolithic hominid populations from each other, thereby implicitly helping discredit the multiregionalist view of the latter as the former's descendants.[75]

Historical "periods" are basically products of our mind,[76] so it is very important not to essentialize our unmistakably conventional systems of periodization. After all, even the Middle Ages and the Renaissance were only identified as distinct "periods" in 1688 and 1855, respectively.[77] Nor, for that matter, was it all that common to view an entire century as a distinct historical unit prior to the 1600s,[78] and even our vision of the decade as a freestanding chunk of history actually dates only to 1931.[79] Indeed, had we normally been counting (and thereby also reckoning the time) in base 9 instead of 10, we would have probably generated fin-de-siècle and millenarian frenzy around the years 1944 (the end of the twenty-fourth 81-year "century") and 1458 (the end of the second 729-year "millennium").[80]

There are many alternative ways to cut up the past, none of which are more natural and hence more valid than others.[81] Any system of periodization is thus inevitably social, since our ability to envision the historical watersheds separating one conventional "period" from another is basically a product of being socialized into specific *traditions* of carving the past. In other words, we need to be mnemonically *socialized* to regard certain historical events as significant "turning points." We thus need to *learn,* for example, to remember "the Reformation" as a process that began with Martin Luther in 1517 (rather than, say, with John Wyclif in the 1370s), and to internalize the distinctly Western mnemonic vision of "the Roman Empire" as a political entity that came to an end in 476 despite the fact that it actually lasted for another 977 years in Byzantium! As jazz fans, we likewise learn to remember João Gilberto and Antonio Carlos Jobim as having actually "pioneered" the bossa nova revolution in 1958 and thereby also implicitly relegate a classic yet highly underrated 1953 recording by Laurindo Almeida and Bud Shank to the dubious status of a mere "forerunner."

Indeed, with the possible exception of the Big Bang, at what point any given stretch of history actually "begins" is never quite self-evident, and there is always more than just a single point that might possibly constitute the formal beginning of a particular historical narrative. After all, even people recounting an event they have just witnessed together (let alone the history of their relationship) often disagree on where their account should begin. In fact, as the pro-life movement keeps reminding us, even the conventional status of birth as an "obvious" biographical point of departure is contestable.[82]

Nor, for that matter, is it all that clear where the story of "human" evolution ought to begin. "Men," after all, "have birthdays, but man does not."[83] And since even the seemingly dramatic evolutionary splits between mollusks and vertebrates or reptiles and mammals were probably not as momentous as we might imagine,[84] would it even be possible to identify the point of transition from manlike apes to apelike men? Should we try, for example, to identify the precise historical moment that marks the branching point of the pongid and hominid lines? Should we perhaps instead try to identify the first hominid species that produced tools? That cooked its food? That acquired an erect posture? That developed language? That produced art?[85]

Consider also the way we mentally organize past military conflicts in conventional units of "war." The so-called Peloponnesian War, for example, may have actually been a conventionally lumped series of several entirely separate conflicts. At the same time, however, one could also argue that it was in fact only a conventionally split part of a much longer conflict between Athens and Sparta, and that the state of nonbelligerency that preceded what we conventionally regard as its "outbreak" in 431 BC was indeed just some brief temporary truce within that conflict.[86] Like the difference between "vacation" and "days off" or "menstruating" and mere "spotting,"[87] the only difference between a merely "temporary" truce and a full-fledged "lasting" peace is the different blocks of time within which they are sociomentally nested.

Along somewhat similar lines, not all Israelis today accept the official national memory of the Arab-Israeli conflict (as manifested, for example, in formal decoration of war veterans by the State) as consisting of five distinct "wars," namely the 1948–49 War of Independence, the 1956 Sinai Campaign, the 1967 Six Day War, the 1973 Yom Kippur War, and the 1982 Lebanon War. As we can see in figure 21, *historical splitters*,[88] for example, also add to this list the 1929 Arab riots, the 1936–39 Arab Revolt, the long series of border incidents and Israeli reprisals from 1953 to 1956, the 1967–70 so-called War of Attrition (during which the total number of Israeli casualties almost exceeded that of the Six Day War),[89] the 1987–93 First Intifada, and the still-ongoing al-Aqsa Intifada. *Historical lumpers*, on the other hand, basically envision a single, essentially continuous Arab-Jewish conflict that has been going on at least since the end of World War I. As Prime Minister Ariel Sharon put it as late as 2001, "the War of Independence is not over yet. 1948 was only one chapter."[90]

Like Winston Churchill's famous epistemic dilemma concerning whether Britain's 1942 victory over the German army in North Africa was just "the end of the beginning" or perhaps "the beginning of the end" of World War II, such taxonomic disputes between lumpers and splitters cannot ever be decisively resolved any more than they can be in zoology. Yet choosing between such competing mnemonic visions is by no means trivial. Killing civilians, for example, has very different moral implications, depending on whether it takes place "during" or "after" a war.

Much of all this depends, of course, on where we locate the "outbreak" of wars. As reflected in the title of a book such as *The Ten Thousand Day War: Vietnam, 1945–1975*,[91] although most of us remember the Vietnam War as having started only in the 1960s, one might also recall a much longer conflict that actually began with the declaration of Vietnam's independence in 1945 without mentally splitting its French and American phases as we normally do.[92] By the same token, although for most Europeans World War II began right after the German invasion of Poland in 1939, for many Americans it only started with the attack on Pearl Harbor two years later, whereas Japanese liberals seem to recall a "Fifteen-Year War" that began with the Japanese occupation of Manchuria in 1931.[93] Indeed, one might

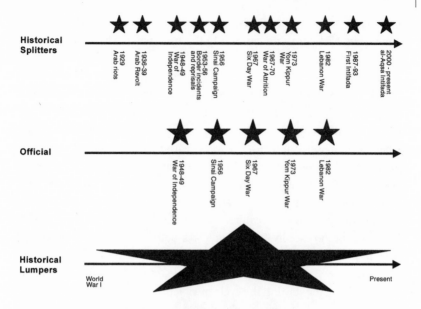

Figure 21 *Israel's Mnemonic Visions of the Arab-Israeli Conflict*

even lump "World War I" and "World War II" together in one's memory as merely two phases of a single conflict which lasted from 1914 to 1945. As one German officer wrote after the French surrender in 1940, "the great battle in France is now ended. It lasted twenty-six years"![94]

Making such mnemonic choices certainly affects the way we normally attribute the actual responsibility for those conflicts. Whether we begin the narrative of the U.S. involvement in Vietnam in 1965 or in 1961, for example, clearly determines whether it is the Johnson or Kennedy administration that we ultimately hold accountable for it. The same applies, of course, to whether we date the actual outbreak of the Second Intifada from Sharon's provocative visit to the Temple Mount on 28 September 2000 or from the violent Palestinian riots protesting his visit the following day.[95]

Such seemingly trivial historiographic differences of opinion often lead to rather heated mnemonic battles somewhat resembling angry disputes between children ("*she* started it, Mom") over the onset of fights. Americans, for example, get extremely annoyed by the rather pervasive Japanese portrayal of the atomic bombings of Hiroshima and Nagasaki as essentially *unprovoked* attacks.[96] At the same time, however, they usually begin the narrative of the Gulf War with the seemingly unprovoked Iraqi invasion of Kuwait in 1990, in marked contrast with the standard Iraqi narrative, which goes back almost a century to the time when Kuwait was still an integral part of Iraq!

In a somewhat similar vein, in sharp contrast to Al-Qaeda leaders, who date the actual outbreak of their current war against the United States from the U.S. cruise missile attacks on their camps in Afghanistan in August 1998[97] (thereby quite conveniently ignoring their prior attacks on two American embassies in Africa two weeks earlier), the United States usually opens the narrative three years later with the infamous 11 September attacks on the World Trade Center and the Pentagon. In fact, that is precisely why it persistently portrays its 2001 campaign in Afghanistan—which U.S. television networks featured under the heading "America Strikes Back"—as pronouncedly "retaliatory." As both Palestinians and Israelis have demonstrated again and again throughout the Second Intifada, by presenting one's acts as a *response* ("revenge," "reprisal"), one essentially puts the blame for starting the cycle of violence on the other side.

Consider also the inevitably unsolvable dilemma inherently involved in any serious effort to offer a fair historical account of the current conflict between Serbs and Albanians in Kosovo. Should such a narrative open, for example, with the Serbian atrocities against Kosovo's Albanians in 1999, or should one maybe try to put those in some "deeper" historical context? And if the latter course is chosen, should the story then begin with Yugoslav president Tito's decision to grant the province autonomy in 1974? With the Serbian takeover of Kosovo in 1912? Should one perhaps go back to the 1683–99 war between Turkey and Austria that led to "the Great Migration" of hundreds of thousands of Serbs from the province in 1690, thereby helping Albanians ultimately become the largest ethnic community there?[98]

As one would expect, Albanians usually begin this narrative sometime between 1690 and 1912, specifically noting that when Serbia conquered Kosovo in 1912 it was essentially an Albanian province. Serbs, on the other hand, prefer either some earlier historical point of "departure" (specifically noting, for example, that prior to "the Great Migration" Kosovo's population was predominantly Serbian) or a much later one that postdates their reconquest of the province in 1912! Though each side in this conflict clearly tends to regard its own narrative as the only correct one, offering a fair historical account may very well require some willingness to actually consider multiple narratives with *multiple beginnings*.

In the Beginnings 5

The special mnemonic status of beginnings is quite evident from the disproportionately high representation, in our general memories from college, of the first few weeks of our freshman year.[1] It also explains the significant role of "origin myths" in defining social communities as well as in solidifying the legitimacy of political regimes.

Origins help articulate identities, and where communities locate their beginnings tells us quite a lot about how they perceive themselves. The frescoes at the Panthéon in Paris featuring the baptism of the Frankish king Clovis following his victory over the Alamanni at the Battle of Tolbiac in 496, for example, are specifically designed to represent the birth of an unmistakably Christian France. The official commemoration of the birth of Mohammed as a national holiday likewise underscores Jordan's and Somalia's identity as distinctly Muslim nation-states.

Indeed, of the 191 countries whose national calendars I have examined, 176 officially celebrate one or more national holidays specifically designed to commemorate their spiritual "origins." Thus, on ten of the eleven days designated on their national calendar as commemorative holidays (Feast of the Immaculate Conception, Christmas, Feast of the Epiphany, Easter Monday, Ascension Day, Corpus Christi, Whit Monday, Saint Stephen's Day, Feast of the Assumption of the Blessed Virgin Mary, and All Saints' Day), Austrians officially commemorate their distinctly Christian origins. The situation is quite similar in India (fourteen of its seventeen national commemorative holidays are specifically designed to celebrate its Hindu, Buddhist, Jainist, Christian, Muslim, and Sikh "roots"), Ethiopia (nine of eleven such days), Spain (ten of twelve), Indonesia (eight of nine), Senegal (eleven of thirteen), and Liechtenstein (all fourteen). In fact, the births of Christianity and Islam are officially commemorated as national holidays (Christmas, Prophet's Birthday) in 149 and 46 countries, respectively.

The social commemoration of "origins" is not confined in any way to nations or religious communities and is just as evident in the various anniversaries through which cities, colleges, and companies celebrate the historic moments when they were founded and couples commemorate their wedding. Indeed, the difference between bicentennial celebrations of national "founding moments"[2] and wedding anniversaries is only a matter of scale.

Antiquity

Note, in this regard, that Austria's and Jordan's Christian and Muslim spiritual origins lie 2,000 and 1,400 years "back," respectively. India's Buddhist, Jain, and Hindu roots, of course, go even "deeper." By the same token, on Passover, Succoth, and Shavuot, Jews essentially commemorate events that allegedly took place some thirty-two centuries ago.

We have already seen earlier that *historical depth* helps widen the span of our collateral genealogical ties. As we shall see now, "deepening" our historical roots also helps solidify our identity as well as legitimacy.

In the same way that taller buildings require deeper foundations, pedigrees assume greater solidity the "deeper" they go back in time. Just as old acquaintances sometimes try to consolidate their tie by reminding each other that they "go back" to high school,[3] the "deeper" one's pedigree, the more commanding one's stature as a descendant. As becomes quite apparent from comparing fourth- and tenth-generation purebred canine or equine champions, the "deeper" it is, the more respectable it looks. Thus, when social status is hereditary, the genealogies of social elites tend to be particularly "deep."[4] This explains the tremendous pride of nations like China, Mexico, and Italy in the *antiquity* of their civilizations.

Consider also the sociomnemonic significance of the publication and subsequent televised serialization of Alex Haley's *Roots* in the 1970s. By opening his aptly titled best-seller with his great-great-great-great-great-great-grandparents back in the Gambia in 1750,[5] Haley certainly revolutionized the way we envision the history of black Americans. Beginning his narrative when their ancestors were still living as free persons in West

Africa, he essentially ended the *mnemonic hegemony* of the traditional Eurocentric view of African Americans as "entering" history only upon becoming relevant to Anglo-Americans as slaves.

No wonder social groups often venerate (and, as so explicitly demonstrated by Lutherans, Bolivians, Marxists, and others, are sometimes even self-consciously named after)[6] the *founding ancestors* from whom they symbolically descend. That is why Japan, for example, commemorates every year on National Foundation Day the accession of its legendary founding emperor Jimmu twenty-six centuries ago, and why the most important Muslim holiday (the Feast of Sacrifice) is associated with Abraham. Moreover, the conventional image of the Abbasid and Fatimid caliphates and the papacy is of essentially uninterrupted lines of succession going all the way "back" to Mohammed's own family and Jesus' own disciple Peter. Such quest for antiquity also explains the attempts made by the last shah of Iran to spin a seemingly seamless 2,500-year symbolic thread linking him to Persia's *first* king, Cyrus,[7] despite the embarrassing fact that the royal Pahlavi "dynasty" actually went back only one generation, to his own father!

The "deeper" we go back in time, of course, the wider our choice of founding ancestors. Like the Mississippi and the Nile, we all have more than just a single genealogical "source." After all, even by going only three generations "back" I can already trace (through my mother's mother's mother, my mother's mother's father, my mother's father's mother, my mother's father's father, my father's mother's mother, my father's mother's father, my father's father's mother, and my father's father's father) no less than eight unmistakably distinct "origins," and their number keeps growing exponentially the "deeper" I go back in time. Given the realities of intermarriage and immigration, such *multiple origins* often entail multiple ethnic and national as well as racial identities.[8] Even a black nationalist like William Du Bois, after all, could trace his origins back to his Dutch, rather than African, "roots."[9]

Priority

Trying to establish "deep" pedigrees might also entail reviving old, sometimes extinct group identities. Evidently inspired by the anticipated collapse of the Ottoman Empire, many nineteenth- and early twentieth-century nationalist movements in southeastern Europe and the Middle

East tried to effectively resuscitate ancient protonational regional identities that had been either actively suppressed or simply forgotten for many centuries of Muslim mnemonic hegemony.[10] Thus, basically downplaying the significance of the Arab and Ottoman conquests of their lands, Egyptian and Greek nationalists tried to emphasize the cultural continuity between ancient and modern Egypt and Greece.[11] In a similar vein, essentially trying to establish unmistakably pre-Muslim national pasts, Turkish nationalists even claimed genealogical ties to Anatolia's Hittite, Phrygian, Trojan, and other ancient inhabitants, while Lebanon's Maronites played up their alleged Phoenician "roots."[12]

As expressed by the way these nationalist movements positioned their protonational "origins" historically vis-à-vis the Ottoman Empire or Islam, antiquity often implies *priority*. Publishing a history of Egypt that not only did not begin but actually ended with the Arab conquest in the seventh century[13] was clearly designed to remind fellow Egyptians that they were there long *before* the Arabs. By the same token, when Spaniards nowadays portray the late-medieval Christian victories over the Moors as a "reconquest," they are implicitly invoking the memory of unmistakably Christian (that is, *pre-Muslim*) early medieval Spain. And when Israeli ultranationalists present their country's occupation of the West Bank in 1967 as a "liberation," it is likewise designed to invoke the memory of an ancient Jewish presence in that region that clearly *preceded* its conquest by the Arabs circa 640.

In short, whether it is Hutu nationalists dismissing the Tutsi as having come to their region *much too recently* to qualify as full members of the Burundian nation,[14] or people with "*old*" money" contemptuously disparaging the so-called *nouveaux* riches, there is often a comparative aspect to the association of antiquity with legitimacy. And in the same way that we use historical priority to support property claims, such as when presenting library carrels or parking spots as ours just because we were there *"first,"* we also consider it a source of greater legitimacy vis-à-vis others.[15]

To appreciate the inevitably relational nature of the way we establish historical priority, note, for example, that despite the fact that Anglo-Americans have already been living in North America for four centuries, their considerable "antiquity" vis-à-vis recent immigrants from Korea or Kenya certainly pales in comparison with Native Americans'. Thus, when the white American in figure 22 points to the Chicano family and arrogantly announces that "it's time to reclaim America from illegal immigrants," he gets a prompt history lesson from the somber-looking Native

Figure 22 *"It's Time to Reclaim America from Illegal Immigrants!"* San Diego Union Tribune, *1994. Steve Kelley, Copley News Service.*

American, who quietly reminds him, "I'll help you pack"! Compared to him, of course, the white American is as much an "illegal immigrant" as the Chicano.

Notice, however, that even the status of "native" is ultimately relational, being essentially a function of having others come to your region after you. After all, if we regard Berbers as the "original" inhabitants of North Africa,[16] it is only because they had arrived there *before the Arabs did.* Although highly evocative labels such as "*Native* Americans" (or its Canadian equivalent, "*First* Nations"), "*Aborigines,*" and "*indigenous* cultures" implicitly portray their bearers as part of the *original* natural landscape of the lands they inhabit, they are actually assigned to them by white people only because they had been living there *prior to the arrival of Europeans!*

Given the way we associate priority with legitimacy, no wonder there are so many mnemonic battles where each side basically tries to *out-past* the other—as if somehow claiming that "my past is longer than yours"—by essentially invoking *earlier* "origins" and thereby implicitly challenging the validity of the beginning of the other side's narrative as an acceptable historical point of departure. Romanian nationalists who claim Roman ancestry thus try to play up the archaeological evidence of ancient Roman settlements in the highly disputed region of Transylvania, in marked contrast with their Hungarian counterparts, who very often quite conveniently fail to even mention those unmistakably pre-Hungarian settlements in their history textbooks.[17] By the same token, essentially challenging the pronouncedly secular Zionist narrative of the modern Jewish settlement of Palestine, Israel's national-religious movement's history textbooks begin the story not with "the first *aliyah*" in 1882 but with a group of Hasidic pilgrims from Belarus who came to Palestine in 1777.[18] The unmistakable political undertones of such mnemonic battles were likewise quite evident when the European Union was planning the first museum specifically designed to depict Europe's "continental" history, and Greece, obviously incensed by a proposal that the exhibit would begin with the (unmistakably Franco-German) ninth-century empire of Charlemagne, emphatically insisted that the "origins" of European civilization be officially pushed thirteen centuries further back, to the classical period.[19]

Similarly, as we can see in figure 23, while Serbs claim that Kosovo was *originally* settled by their ancestors, the southern Slavs, in the sixth century (that is, long before its Ottoman-induced Albanization after their "Great Migration" in 1690), Albanians like to note that when the southern Slavs first came to the province it was already inhabited by *their* ancestors, ancient Illyrian tribes who had lived there for many centuries *before* them.[20] By the same token, as we can see in figure 24, while Arabs essentially regard Israelis as usurpers who have come to "Palestine" only *very recently*, Israelis keep stressing the Jewish presence in "the Land of Israel" long before its conquest by the Arabs in the seventh century.[21] Trying to out-past them, Palestinians go still further "back," playing up their even *earlier* Philistine roots. Challenging Jewish claims to the highly contested city of Jerusalem, they likewise claim descent from the ancient Jebusites, who had lived there, according to the Bible, *prior to* its conquest by King David three thousand years ago,[22] just as they do from other "indigenous" Canaanite peoples who had inhabited that region prior to its conquest by Joshua a couple of centuries earlier. Many Jews, of course, consider such claims somewhat trivial given God's alleged pledge to Abraham even earlier regarding "the Promised Land"!

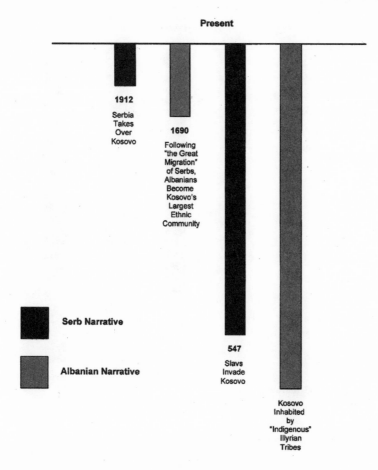

Figure 23 *Serb and Albanian Historical Claims over Kosovo*

The politics of historical priority also offers us some further insight into the logic of mnemonic decapitation, since the quest for such priority is occasionally offset by political expediency. After all, given the greater historical "depth" of the Albanian narrative, for example, it certainly makes no sense for Serbs to even try to extend their memories of Kosovo beyond the sixth century. That may also explain the otherwise peculiar choice of 1840 as the actual historical point of departure on a public, semiofficial Israeli plaque statistically narrating the demographic history of Jerusalem, as it quite conveniently happens to be the first time in census-documented his-

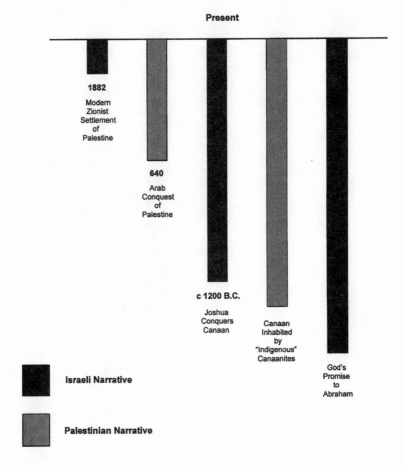

Present

1882
Modern Zionist Settlement of Palestine

640
Arab Conquest of Palestine

c 1200 B.C.
Joshua Conquers Canaan

Canaan Inhabited by "Indigenous" Canaanites

God's Promise to Abraham

Israeli Narrative

Palestinian Narrative

Figure 24 Arab and Jewish Historical Claims over Palestine

tory when Jews finally surpassed both Muslims and Christians to become the disputed city's largest religious community.

Figures 23 and 24 also remind us that antiquity is inherently relative and that the meaning of how long ago any given historical event happened is usually a function of its temporal location *in relation to other events*. Thus, having dismissed a Palestinian allusion to "the Haram al-Sharif, called the Temple Mount by Israelis"[23] by claiming that since "the first construction . . . on this site was the ancient Israelite temple . . . the Hebrew expression Har ha-Bayit . . . deserves priority," a Jewish nationalist then adds that "*only . . . in the seventh century* was the term Haram al-Sharif applied to this site."[24] As his clearly dismissive use of the word *only* indicates, considerable historical distances are often trivialized as part of the process of out-pasting. Mocking the very idea of celebrating the two-hundredth anniversary of the British settlement of Australia in 1988, some Aborigines thus bitterly quipped that "40,000 years don't make a Bicentennial."[25]

To fully understand the mnemonic significance of historical objects and events, we thus need to consult the actual "time maps" that help situate them within socially meaningful historical contexts. The antiquity of a 700-year-old mosque on the West Bank, for example, looks considerably less awesome on the mnemonic "map" of Jewish settlers who claim that, having been originally constructed two thousand years ago as a site for Jewish worship, it has in fact functioned as a mosque "*only* since 1267."[26] The historical weight of more than fourteen centuries of Serb presence in Kosovo is likewise significantly pared down on Albanian "maps" that quite understandably venture in their retrospection well beyond the sixth century.

It is virtually impossible to make any sense of the mnemonic battles over Kosovo, Jerusalem, or the settlement of America and Australia without constantly referring to such "time maps" as general orientational guides, since the meaning of historical objects and events is inevitably tied to these exceptionally critical mental structures within which they are sociomnemonically situated. Furthermore, to fully appreciate the nuanced complexity of such battles, we must also bear in mind that there is usually *more than just one* such "map" we need to consult at any given point!

As we very well know, each of the different parties waging such heated mnemonic battles tends to regard its own historical narrative, which is normally based on its own typically one-sided "time maps," as the only correct one, which is quite understandable given the unmistakably partisan political agenda it is specifically designed to promote. A more dispassionate, nonpartisan, and therefore impartial historical account would require some willingness to consider *multiple narratives*, which inevitably imply the possibility of entertaining multiple perspectives on the past.

Unlike the somewhat nihilistic "postmodern" critique of historical positivism, however, such a pronouncedly *pluralistic view of history* does not necessarily imply disregarding the veracity of what is being remembered, since the main object of its criticism is not historical objectivity per se but the common tendency to presume a single perspective on the past. After all, there is no good reason not to assume that *both* the Palestinian and Israeli narratives of the history of Jerusalem are to a large extent factually accurate, as are the Serb and Albanian narratives of the history of Kosovo and the Icelandic and Spanish accounts of the European "discovery" of America. What is problematic about being historically biased is not only the deliberate fabrication, distortion, or omission of actual facts but also the pronouncedly partisan, politically motivated mnemonic selectivity that leads one to dismiss or ignore any historical narrative other than one's own.

What *both* parties in such conflictual situations tend to remember from the past is primarily (though not always exclusively) based on fact, and the heated mnemonic battles between them usually revolve around the specific manner in which those facts are respectively registered on their unmistakably different, often competing "time maps." As I have demonstrated throughout this book, there are not only many different patterns of organizing the past in our heads but also various different methods for arranging each of those specific patterns. Only a pronouncedly multiperspectival look[27] at several such "maps" *together* can provide us with a complete picture of the inevitably multilayered, multifaceted social topography of the past.

Notes

Introduction

1. Jan Assmann, *Moses the Egyptian: The Memory of Egypt in Western Monotheism* (Cambridge, Mass.: Harvard University Press, 1997), p. 11.
2. Eviatar Zerubavel, *Social Mindscapes: An Invitation to Cognitive Sociology* (Cambridge, Mass.: Harvard University Press, 1997), pp. 97–99.
3. Marshall Sahlins, *Historical Metaphors and Mythical Realities: Structure in the Early History of the Sandwich Islands Kingdom* (Ann Arbor: University of Michigan Press, 1981), pp. 13–14.
4. Stephen Howe, *Afrocentrism: Mythical Pasts and Imagined Homes* (London: Verso, 1998), p. 105. Emphasis added.
5. See also Robert E. Park and Ernest W. Burgess, *Introduction to the Science of Sociology*, abridged ed. (Chicago: University of Chicago Press, 1969), pp. 360–61, 365.
6. Michael Frisch, "American History and the Structures of Collective Memory: A Modest Exercise in Empirical Iconography," *Journal of American History* 75 (1989): 1130–55.
7. E. Zerubavel, *Social Mindscapes*, pp. 17–18, 90. See also Iwona Irwin-Zarecka, *Frames of Remembrance: The Dynamics of Collective Memory* (New Brunswick, N.J.: Transaction, 1994), p. 47 on "communities of memory."
8. Jan Vansina, *Oral Tradition as History* (Madison: University of Wisconsin Press, 1985), p. 149.
9. E. Zerubavel, *Social Mindscapes*, p. 97.
10. Roger C. Schank and Robert P. Abelson, "Scripts, Plans, and Knowledge," in *Thinking: Readings in Cognitive Science*, edited by P. N. Johnson-Laird and P. C. Wason (Cambridge: Cambridge University Press, 1977), p. 430.
11. Margaret S. Steffensen, Chitra Joag-Dev, and Richard C. Anderson, "A Cross-Cultural Perspective on Reading Comprehension," *Reading Research Quarterly* 15 (1979): 10–29.
12. E. Zerubavel, *Social Mindscapes*, pp. 87–91.
13. Frederic C. Bartlett, *Remembering: A Study in Experimental and Social Psychology* (Cam-

bridge: Cambridge University Press, 1932); Walter Kintsch and Edith Greene, "The Role of Culture-Specific Schemata in the Comprehension and Recall of Stories," *Discourse Processes* 1 (1978): 1–13; Marjorie Y. Lipson, "The Influence of Religious Affiliation on Children's Memory for Text Information," *Reading Research Quarterly* 18 (1983): 448–57; Robert Pritchard, "The Effects of Cultural Schemata on Reading Processing Strategies," *Reading Research Quarterly* 25 (1990): 273–95.

14. Hayden White, "The Historical Text as Literary Artifact," in *Tropics of Discourse: Essays in Cultural Criticism* (Baltimore: Johns Hopkins University Press, 1978), pp. 81–99.

15. Yael Zerubavel, *Recovered Roots: Collective Memory and the Making of Israeli National Tradition* (Chicago: University of Chicago Press, 1995), pp. 17–22.

16. E. Zerubavel, *Social Mindscapes*, pp. 13, 84.

17. Ibid., p. 87.

18. Wendy Traas, "Turning Points and Defining Moments: An Exploration of the Narrative Styles That Structure the Personal and Group Identities of Born-Again Christians and Gays and Lesbians" (unpublished manuscript, Rutgers University, Department of Sociology, 2000); Jenna Howard, "Memory Reconstruction in Autobiographical Narrative Construction: Analysis of the Alcoholics Anonymous Recovery Narrative" (unpublished manuscript, Rutgers University, Department of Sociology, 2000).

19. See also Yosef H. Yerushalmi, *Zakhor: Jewish History and Jewish Memory* (Seattle: University of Washington Press, 1982).

20. William Hirst and David Manier, "Remembering as Communication: A Family Recounts Its Past," in *Remembering Our Past: Studies in Autobiographical Memory*, edited by David C. Rubin (Cambridge: Cambridge University Press, 1996), pp. 271–88.

21. Robyn Fivush, Catherine Haden, and Elaine Reese, "Remembering, Recounting, and Reminiscing: The Development of Autobiographical Memory in Social Context," in *Remembering Our Past: Studies in Autobiographical Memory*, edited by David C. Rubin (Cambridge: Cambridge University Press, 1996), pp. 341–58.

22. See also E. Zerubavel, *Social Mindscapes*, p. 16.

23. See, for example, M. T. Clanchy, *From Memory to Written Record: England, 1066–1307* (Cambridge, Mass.: Harvard University Press, 1979), p. 202.

24. Eviatar Zerubavel, *Patterns of Time in Hospital Life: A Sociological Perspective* (Chicago: University of Chicago Press, 1979), pp. 45–46.

25. See also Georg Simmel, "Written Communication," in *The Sociology of Georg Simmel*, edited by Kurt H. Wolff (New York: Free Press, 1950 [1908]), pp. 352–55 (page citations are to the reprint edition); Max Weber, *Economy and Society: An Outline of Interpretive Sociology* (Berkeley and Los Angeles: University of California Press, 1978 [1925]), pp. 219, 957 (page citations are to the reprint edition); Clanchy, *From Memory to Written Record;* Edward Shils, *Tradition* (Chicago: University of Chicago Press, 1981), pp. 109–12, 120–24, 140–47.

26. Pierre Nora, "Between Memory and History: Les Lieux de Memoire," *Representations* 26 (1989): 7–25.

27. See, for example, Y. Zerubavel, *Recovered Roots.*

28. See also Henry Glassie, *Passing the Time in Ballymenone: Culture and History of an Ulster Community* (Philadelphia: University of Pennsylvania Press, 1982), pp. 619–65.

29. Ferdinand de Saussure, *Course in General Linguistics* (New York: Philosophical Library, 1959), pp. 115–22. See also E. Zerubavel, *Social Mindscapes*, pp. 72–76.

30. For a classic sample of the possible products of such a "formal" sociological methodology, see Simmel, *The Sociology of Georg Simmel*, pp. 21–23, 40–57, 87–408. See also Allan V. Horwitz, *The Logic of Social Control* (New York: Plenum, 1990); Eviatar Zerubavel, *The Fine Line: Making Distinctions in Everyday Life* (New York: Free Press, 1991); Kristen Purcell, "Leveling the Playing Field: Constructing Parity in the Modern World" (Ph.D. diss., Rutgers University, 2001).

31. See also Eviatar Zerubavel, "If Simmel Were a Fieldworker: On Formal Sociological Theory and Analytical Field Research," *Symbolic Interaction* 3, no. 2 (1980): 25–33.

32. For some notable exceptions, however, see Maurice Halbwachs, *The Social Frameworks of Memory*, in *Maurice Halbwachs on Collective Memory*, edited by Lewis A. Coser (Chicago: University of Chicago Press, 1992 [1925]), pp. 37–189 (page citations are to the reprint edition); Shils, *Tradition*; Paul Connerton, *How Societies Remember* (Cambridge: Cambridge University Press, 1989).

33. On "cognitive universalism," see E. Zerubavel, *Social Mindscapes*, pp. 2–6, 20–22.

Chapter One

1. Alistair Horne, *To Lose a Battle: France 1940* (Boston: Little, Brown & Co., 1969), p. 582. Emphasis added.

2. Ibid.

3. See also Peter J. Bowler, *Theories of Human Evolution: A Century of Debate, 1844–1944* (Baltimore: Johns Hopkins University Press, 1986), p. 13.

4. Ferdinand Saussure, *Course in General Linguistics*, (New York: Philosophical Library, 1959), pp. 115–22. See also Eviatar Zerubavel, *Social Mindscapes: An Invitation to Cognitive Sociology* (Cambridge, Mass.: Harvard University Press, 1997), pp. 72–76.

5. *New York Times*, 21 January 2001, sec. A, p. 13.

6. See also Michael A. Bernstein, *Foregone Conclusions: Against Apocalyptic History* (Berkeley and Los Angeles: University of California Press, 1994).

7. Hayden White, "The Historical Text as Literary Artifact," in *Tropics of Discourse: Essays in Cultural Criticism* (Baltimore: Johns Hopkins University Press, 1978), p. 83.

8. Ibid., p. 87.

9. See also Peter L. Berger, *Invitation to Sociology: A Humanistic Perspective* (Garden City, N.Y.: Doubleday Anchor, 1963), pp. 56–65.

10. White, "The Historical Text as Literary Artifact." See also Misia Landau, "Human Evolution as Narrative," *American Scientist* 72 (1984): 262–68; Bowler, *Theories of Human Evolution*, pp. 13–14.

11. White, "The Historical Text as Literary Artifact." See also Jean M. Mandler, *Stories, Scripts, and Scenes: Aspects of Schema Theory* (Hillsdale, N.J.: Lawrence Erlbaum, 1984), p. 18; S. Wojciech Sokolowski, "Historical Tradition in the Service of Ideology," *Conjecture* (September 1992): 4–11; Yael Zerubavel, *Recovered Roots: Collective Memory and the Making of Israeli National Tradition*, pp. 216–21.

12. See also Henry Glassie, *Passing the Time in Ballymenone: Culture and History of an Ulster Community* (Philadelphia: University of Pennsylvania Press, 1982), p. 651.

13. Agnes Hankiss, "Ontologies of the Self: On the Mythological Rearranging of One's Life-History," in *Biography and Society: The Life History Approach in the Social Sciences*, edited by Daniel Bertaux (Beverly Hills, Calif.: Sage, 1981), pp. 205–7; Dan P.

McAdams, *The Stories We Live By: Personal Myths and the Making of the Self* (New York: William Morrow, 1993), pp. 104–5.

14. Rulon S. Wells, "The Life and Growth of Language: Metaphors in Biology and Linguistics," in *Biological Metaphor and Cladistic Classification: An Interdisciplinary Perspective*, edited by Henry M. Hoenigswald and Linda F. Wiener (Philadelphia: University of Pennsylvania Press, 1987), pp. 51–52; Peter J. Bowler, *The Invention of Progress: The Victorians and the Past* (Oxford: Basil Blackwell, 1989), p. 10; Gordon Graham, *The Shape of the Past: A Philosophical Approach to History* (Oxford: Oxford University Press, 1997), pp. 47, 65.

15. Bowler, *Theories of Human Evolution*, pp. 50–52.

16. See Barry Schwartz, *Vertical Classification: A Study in Structuralism and the Sociology of Knowledge* (Chicago: University of Chicago Press, 1981), pp. 79–93.

17. Jacob Bronowski, *The Ascent of Man* (Boston: Little, Brown & Co., 1973).

18. George W. Stocking, "The Dark-Skinned Savage: The Image of Primitive Man in Evolutionary Anthropology," in *Race, Culture, and Evolution: Essays in the History of Anthropology* (New York: Free Press, 1968), p. 118; Bowler, *The Invention of Progress*, p. 10.

19. Christopher Lasch, *The True and Only Heaven: Progress and Its Critics* (New York: W. W. Norton, 1991).

20. See also Lewis A. Coser and Rose L. Coser, "Time Perspective and Social Structure," in *Modern Sociology: An Introduction to the Science of Human Interaction*, edited by Alvin W. Gouldner and Helen P. Gouldner (New York: Harcourt, Brace & World, 1963), pp. 638–47.

21. Bowler, *The Invention of Progress*; Arthur Herman, *The Idea of Decline in Western History* (New York: Free Press, 1997), pp. 25, 30–36.

22. See also Herman, *The Idea of Decline in Western History*; Graham, *The Shape of the Past*, pp. 83–111.

23. Wells, "The Life and Growth of Language," pp. 51–52.

24. Graham, *The Shape of the Past*, p. 83.

25. See also *The Babylonian Talmud* (London: Soncino, 1938), vol. 2, pt. 2, p. 549 (*Tractate Shabbath*, 112b).

26. Konrad Lorenz, *On Aggression* (New York: Bantam, 1971), p. 217.

27. Herman, *The Idea of Decline in Western History*, p. 13.

28. Fred Davis, *Yearning for Yesterday: A Sociology of Nostalgia* (New York: Free Press, 1979), pp. 104–8.

29. Bowler, *The Invention of Progress*, pp. 195–96; Peter J. Bowler, *Life's Splendid Drama: Evolutionary Biology and the Reconstruction of Life's Ancestry, 1860–1940* (Chicago: University of Chicago Press, 1996), pp. 431–32; Herman, *The Idea of Decline in Western History*, pp. 91–127, 225–89.

30. Wendy Traas, "Turning Points and Defining Moments: An Exploration of the Narrative Styles That Structure the Personal and Group Identities of Born-Again Christians and Gays and Lesbians" (unpublished manuscript, Rutgers University, Department of Sociology, 2000).

31. Jenna Howard, "Memory Reconstruction in Autobiographical Narrative Construction: Analysis of the Alcoholics Anonymous Recovery Narrative" (unpublished manuscript, Rutgers University, Department of Sociology, 2000).

32. Andrew Abbott, "On the Concept of the Turning Point," *Comparative Social Research* 16 (1997): 85–105.

33. Tamara K. Hareven and Kanji Masaoka, "Turning Points and Transitions: Perceptions of the Life Course," *Journal of Family History* 13 (1988): 272.

34. John Gribbin and Jeremy Cherfas, *The Monkey Puzzle: Reshaping the Evolutionary Tree* (New York: Pantheon, 1982), p. 45; Bowler, *The Invention of Progress*, pp. 12–13; Bowler, *Life's Splendid Drama*, pp. 16, 424–25.

35. Johannes Fabian, *Time and the Other: How Anthropology Makes Its Object* (New York: Columbia University Press, 1983); Bowler, *Theories of Human Evolution*, pp. 41, 50; Clive Gamble, *Timewalkers: The Prehistory of Global Colonization* (Cambridge: Mass.: Harvard University Press, 1994), pp. 17–18.

36. Bowler, *Theories of Human Evolution*, pp. 210–23. See also Landau, "Human Evolution as Narrative."

37. Bowler, *Life's Splendid Drama*, pp. 314, 339–52.

38. Auguste Comte, *Cours de Philosophie Positive*, in *Auguste Comte and Positivism: The Essential Writings*, edited by Gertrud Lenzer (New York: Harper Torchbooks, 1975), p. 230. See also p. 285.

39. See, for example, ibid., pp. 199, 211, 231, 233, 285.

40. Bowler, *The Invention of Progress*, p. 10.

41. József Böröcz, "Sticky Features: Narrating a Single Direction" (paper presented at the "Beginnings and Endings" seminar at the Center for the Critical Analysis of Contemporary Culture, Rutgers University, New Brunswick, September 1999).

42. Bowler, *Theories of Human Evolution*, pp. 214–15, 218–21; Bowler, *Life's Splendid Drama*, pp. 424–25.

43. Nancy Stepan, *The Idea of Race in Science: Great Britain 1800–1960* (Hamden, Conn.: Archon Books, 1982), pp. 11–13; Bowler, *Life's Splendid Drama*, p. 46.

44. Michael Hammond, "The Expulsion of the Neanderthals from Human Ancestry: Marcellin Boule and the Social Context of Scientific Research," *Social Studies of Science* 12 (1982): 5, 20.

45. Ibid., pp. 2, 29; Bowler, *The Invention of Progress*, p. 126; Bowler, *Life's Splendid Drama*, p. 316.

46. Ian Tattersall and Jeffrey H. Schwartz, *Extinct Humans* (Boulder, Colo.: Westview, 2000), p. 174. See also p. 33.

47. Matt Cartmill, " 'Four Legs Good, Two Legs Bad': Man's Place (if Any) in Nature," *Natural History* 92 (November 1983): 65.

48. Gribbin and Cherfas, *The Monkey Puzzle*, p. 45. See also Bowler, *Theories of Human Evolution*, pp. 14, 41.

49. Konrad Koerner, "On Schleicher and Trees," in *Biological Metaphor and Cladistic Classification: An Interdisciplinary Perspective*, edited by Henry M. Hoenigswald and Linda F. Wiener (Philadelphia: University of Pennsylvania Press, 1987), pp. 111–12. See also Henry M. Hoenigswald, "Language Family Trees, Topological and Metrical," in *Biological Metaphor and Cladistic Classification: An Interdisciplinary Perspective*, edited by Henry M. Hoenigswald and Linda F. Wiener (Philadelphia: University of Pennsylvania Press, 1987), pp. 257–67; Bowler, *The Invention of Progress*, p. 151.

50. Charles Darwin, *The Origin of Species* (New York: Mentor Books, 1958 [1859]), pp. 115–

17, 392–93 (page citations are to the reprint edition); Wells, "The Life and Growth of Language," p. 50.

51. Stephen J. Gould, *Wonderful Life: The Burgess Shale and the Nature of History* (New York: W. W. Norton, 1989), p. 35. See also Gribbin and Cherfas, *The Monkey Puzzle*, p. 45; Tattersall and Schwartz, *Extinct Humans*, p. 33.

52. Gould, *Wonderful Life*, p. 45.

53. Bowler, *Life's Splendid Drama*, p. 426.

54. See, for example, the illustrations in Gould, *Wonderful Life*, pp. 29–35.

55. See also ibid., p. 29.

56. Fabian, *Time and the Other*.

57. See, for example, Eviatar Zerubavel, *The Clockwork Muse: A Practical Guide to Writing Theses, Dissertations, and Books* (Cambridge, Mass.: Harvard University Press, 1999), pp. 40, 46, 83–84.

58. Eviatar Zerubavel, *Patterns of Time in Hospital Life: A Sociological Perspective* (Chicago: University of Chicago Press, 1979), pp. 1–2, 37–38; Eviatar Zerubavel, *The Seven-Day Circle: The History and Meaning of the Week* (New York: Free Press, 1985), pp. 83–84. See also Edmund Leach, "Two Essays concerning the Symbolic Representation of Time," in *Rethinking Anthropology* (London: Athlone, 1961), pp. 124–36; Johanna E. Foster, "Menstrual Time: The Sociocognitive Mapping of 'The Menstrual Cycle,'" *Sociological Forum* 11 (1996): 529–32.

59. Samuel C. Heilman, *The People of the Book: Drama, Fellowship, and Religion* (Chicago: University of Chicago Press, 1983), p. 65.

60. See also David Lowenthal, *The Past Is a Foreign Country* (Cambridge: Cambridge University Press, 1985), p. 241.

61. Mircea Eliade, *The Sacred and the Profane: The Nature of Religion* (New York: Harcourt, Brace & World, 1959), pp. 68–113. See also Marshall Sahlins, *Historical Metaphors and Mythical Realities: Structure in the Early History of the Sandwich Islands Kingdom* (Ann Arbor: University of Michigan Press, 1981), p. 14.

62. David H. Fischer, *The Great Wave: Price Revolutions and the Rhythm of History* (New York: Oxford University Press, 1996), p. 3.

63. Jan Vansina, *Oral Tradition as History* (Madison: University of Wisconsin Press, 1985), p. 21.

64. Ibid., p. 166.

65. On the process of typification, see Alfred Schutz and Thomas Luckmann, *The Structures of the Life-World* (Evanston, Ill.: Northwestern University Press, 1973), pp. 77, 229–41; Peter L. Berger and Thomas Luckmann, *The Social Construction of Reality: A Treatise in the Sociology of Knowledge* (Garden City, N.Y.: Doubleday, 1966), pp. 30–34.

66. Eviatar Zerubavel, "Lumping and Splitting: Notes on Social Classification," *Sociological Forum* 11 (1996): 430.

67. Y. Zerubavel, *Recovered Roots*, pp. 220–21.

68. Eviatar Zerubavel, *Hidden Rhythms: Schedules and Calendars in Social Life* (Chicago: University of Chicago Press, 1981), pp. 59–64.

69. See Michael G. Flaherty, *A Watched Pot: How We Experience Time* (New York: New York University Press, 1999).

70. E. Zerubavel, *Patterns of Time in Hospital Life*, pp. 113–14.

71. Henri Hubert, "Etude Sommaire de la Représentation du Temps dans la Religion et la

Magie," in *Mélanges d'Histoire des Religions*, edited by Henri Hubert and Marcel Mauss (Paris: Félix Alcan and Guillaumin, 1909), p. 207; Pitirim A. Sorokin, *Sociocultural Causality, Space, Time: A Study of Referential Principles of Sociology and Social Science* (Durham, N.C.: Duke University Press, 1943), p. 184.

72. Hubert, "Etude Sommaire de la Représentation du Temps," pp. 197–203; Sorokin, *Sociocultural Causality, Space, Time*, pp. 183–84.

73. Henri Bergson, *Time and Free Will: An Essay on the Immediate Data of Consciousness* (New York: Harper and Row, 1960), pp. 90–128, 222–40.

74. Eviatar Zerubavel, "The Social Marking of the Past: Toward a Socio-Semiotics of Memory," in *The Cultural Turn*, edited by Roger Friedland and John Mohr (Cambridge: Cambridge University Press, in press). See also Emile Durkheim, *The Elementary Forms of Religious Life* (New York: Free Press, 1995 [1912]), p. 313 (page citations are to the reprint edition); Leach, "Two Essays concerning the Symbolic Representation of Time"; W. Lloyd Warner, *The Family of God* (New Haven, Conn.: Yale University Press, 1961), pp. 345–73; Foster, "Menstrual Time," pp. 525–28.

75. E. Zerubavel, *The Seven-Day Circle*, pp. 113–20.

76. Sorokin, *Sociocultural Causality, Space, Time*, p. 212; George Kubler, *The Shape of Time* (New Haven, Conn.: Yale University Press, 1962), p. 75.

77. See also Glassie, *Passing the Time in Ballymenone*, p. 659.

78. A. M. Snodgrass, *The Dark Age of Greece: An Archeological Survey of the Eleventh to the Eighth Centuries B.C.* (Edinburgh: Edinburgh University Press, 1971); Peter James, *Centuries of Darkness: A Challenge to the Conventional Chronology of Old World Archaeology* (New Brunswick, N.J.: Rutgers University Press, 1993); Ian Morris, "Periodization and the Heroes: Inventing a Dark Age," in *Inventing Ancient Culture: Historicism, Periodization, and the Ancient World*, edited by Mark Golden and Peter Toohey (London: Routledge, 1997), pp. 96–131.

79. See also Eviatar Zerubavel, "Language and Memory: 'Pre-Columbian' America and the Social Logic of Periodization," *Social Research* 65 (1998): 328.

80. See also Claude Lévi-Strauss, *The Savage Mind* (Chicago: University of Chicago Press, 1966), p. 257.

81. See also E. Zerubavel, *Social Mindscapes*, pp. 50–52.

82. See, for example, Snodgrass, *The Dark Age of Greece*, pp. 16, 20.

83. Glassie, *Passing the Time in Ballymenone*, pp. 621–22; Nadia Abu El-Haj, *Facts on the Ground: Archaeological Practice and Territorial Self-Fashioning in Israeli Society* (Chicago: University of Chicago Press, 2001), pp. 148–58.

84. Lévi-Strauss, *The Savage Mind*, p 259. Emphasis added.

85. 2 Chronicles, 1:1–9:31, 24:1–27, 29:1–33:20.

86. Clifton Daniel, *Chronicle of America* (Mount Kisco, N.Y.: Chronicle Publications, 1989), pp. 82–105, 142–65.

87. Ulrich Herbert, "Good Times, Bad Times," *History Today* 36 (February 1986): 44. See also Howard Schuman and Jacqueline Scott, "Generations and Collective Memories," *American Sociological Review* 54 (1989): 359–81.

88. E. Zerubavel, "The Social Marking of the Past."

89. Durkheim, *The Elementary Forms of Religious Life*, pp. 303–417.

90. Barry Schwartz, "The Social Context of Commemoration: A Study in Collective Memory," *Social Forces* 61 (1982): 377.

91. Y. Zerubavel, *Recovered Roots*. See also Paul Connerton, *How Societies Remember* (Cambridge: Cambridge University Press, 1989).

92. Maoz Azaryahu, "The Purge of Bismarck and Saladin: The Renaming of Streets in East Berlin and Haifa," *Poetics Today* 13 (1992): 351–66.

93. W. Lloyd Warner, *The Living and the Dead* (New Haven, Conn.: Yale University Press, 1959), pp. 129–30.

94. Ibid., p. 133.

95. Schwartz, "The Social Context of Commemoration," pp. 381–82.

96. The 191-country national holiday data set on which I draw here is based on information gathered from *Europa World Year Book 1997* (London: Europa Publications, 1997); Miranda Haines, ed., *The Traveler's Handbook*, 7th ed. (London: Wexas, 1997); Helene Henderson and Sue Ellen Thompson, eds., *Holidays, Festivals, and Celebrations of the World Dictionary*, 2d ed. (Detroit: Omnigraphics Inc., 1997); Robert S. Weaver, *International Holidays: 204 Countries from 1994 through 2015* (Jefferson, N.C.: McFarland, 1995); Ruth W. Gregory, *Anniversaries and Holidays*, 4th ed. (Chicago: American Library Association, 1983); *Chase's 1997 Calendar of Events* (Chicago: Contemporary Publishing Co., 1996); and various country-specific travel guides.

97. Eviatar Zerubavel, "Calendars and History: A Comparative Study of the Social Organization of National Memory," in *States of Memory: Conflicts, Continuities, and Transformations in National Commemoration*, edited by Jeffrey K. Olick (Durham, N.C.: Duke University Press, in press).

98. G. I. Jones, "Time and Oral Tradition with Special Reference to Eastern Nigeria," *Journal of African History* 6 (1965): 153–55; David P. Henige, *The Chronology of Oral Tradition: Quest for a Chimera* (London: Oxford University Press, 1974), p. 27; Joseph C. Miller, "Introduction: Listening for the African Past," in *The African Past Speaks: Essays on Oral Tradition and History* (Folkestone, England: William Dawson, 1980), pp. 36–37; Vansina, *Oral Tradition as History*, pp. 23–24, 168–69; Mary Douglas, *How Institutions Think* (Syracuse, N.Y.: Syracuse University Press, 1986), pp. 72–73.

99. I quite intentionally do not distinguish here "historical" events from "mythical" or "legendary" ones and am therefore not concerned whether or not any particular commemorated event actually happened, so long as it is conventionally anchored in some collectively shared past.

100. The calendrical data presented here reflect the situation in 1999. Such commemograms, however, are inevitably fluid, since new holidays are added to the calendar and old ones removed from it as countries' political situations change. See E. Zerubavel, "Calendars and History."

101. Hans U. Gumbrecht, *In 1926: Living at the Edge of Time* (Cambridge, Mass.: Harvard University Press, 1997).

102. Eviatar Zerubavel, *The Fine Line: Making Distinctions in Everyday Life* (New York: Free Press, 1991), pp. 9–10, 23–24, 27, 30–31, 72. See also Richard Sorabji, *Time, Creation, and the Continuum: Theories in Antiquity and the Early Middle Ages* (Ithaca, N.Y.: Cornell University Press, 1983).

103. See also E. Zerubavel, *Patterns of Time in Hospital Life*, pp. 9–11.

104. Niles Eldredge and Stephen J. Gould, "Punctuated Equilibria: An Alternative to Phyletic Gradualism," in *Models in Paleobiology*, edited by Thomas J. Schopf (San Francisco: Freeman, Cooper, & Co., 1972), p. 97; Gribbin and Cherfas, *The Monkey*

Puzzle, p. 56; Bowler, *The Invention of Progress*, pp. 147–48; Gamble, *Timewalkers*, p. 76.

105. Arthur O. Lovejoy, *The Great Chain of Being: A Study of the History of an Idea* (Cambridge, Mass.: Harvard University Press, 1936), pp. 242–87.

106. Darwin, *The Origin of Species*, p. 435. See also Richard Dawkins, *River out of Eden: A Darwinian View of Life* (New York: Basic Books, 1995), pp. 83–84.

107. Eldredge and Gould, "Punctuated Equilibria," p. 89; Dawkins, *River out of Eden*, p. 93.

108. Eldredge and Gould, "Punctuated Equilibria," p. 87; Bowler, *Life's Splendid Drama*, p. 353. See also Lovejoy, *The Great Chain of Being*, pp. 231–36.

109. Eldredge and Gould, "Punctuated Equilibria," pp. 84, 96, 98; Stephen J. Gould, *The Structure of Evolutionary Theory* (Cambridge, Mass.: Harvard University Press, 2002), pp. 745–1024.

110. John Reader, *Missing Links: The Hunt for Earliest Man* (Boston: Little, Brown, & Co., 1981), p. 204; Gamble, *Timewalkers*, p. 76.

Chapter Two

1. See also Agnes Hankiss, "Ontologies of the Self: On the Mythological Rearranging of One's Life-History," in *Biography and Society: The Life History Approach in the Social Sciences*, edited by Daniel Bertaux (Beverly Hills, Calif.: Sage, 1981) pp. 205, 208–9.

2. See also József Böröcz, "Sticky Features: Narrating a Single Direction" (paper presented at the "Beginnings and Endings" seminar, Center for the Critical Analysis of Contemporary Culture, Rutgers University, New Brunswick, N.J., September 1999).

3. See also Herbert Spencer, *Principles of Sociology* (Hamden, Conn.: Archon, 1969 [1876]), pp. 444–47 (page citations are to the reprint edition); George Kubler, *The Shape of Time* (New Haven, Conn.: Yale University Press, 1962), pp. 56, 72; Edward Shils, *Tradition* (Chicago: University of Chicago Press, 1981), pp. 34–54.

4. See also Alfred Schutz, "Phenomenology and the Social Sciences," in *Collected Papers, vol. 1: The Problem of Social Reality* (The Hague: Martinus Nijhoff, 1973), p. 136.

5. Karl Marx, "The Eighteenth Brumaire of Louis Bonaparte," in *The Marx-Engels Reader*, 2d ed., edited by Robert C. Tucker (New York: W. W. Norton, 1978), p. 595.

6. *New York Times*, 2 August 1994, International section, sec. A, p. 2.

7. Maureen Dowd, "Center Holding," *New York Times*, 20 May 1998, sec. A, p. 23.

8. James Brooke, "Conquistador Statue Stirs Hispanic Pride and Indian Rage," *New York Times*, 9 February 1998, sec. A, p. 10.

9. See Edwin A. Abbott, *Flatland: A Romance of Many Dimensions* (New York: Dover, 1992).

10. Helen R. F. Ebaugh, *Becoming an Ex: The Process of Role Exit* (Chicago: University of Chicago Press, 1988), pp. 156, 162. See also pp. 173–80.

11. Alvin Toffler, *Future Shock* (New York: Random House, 1970).

12. Vance Packard, *The Waste Makers* (New York: David McKay, 1960).

13. Fred Davis, *Yearning for Yesterday: A Sociology of Nostalgia* (New York: Free Press, 1979); David Lowenthal, *The Past Is a Foreign Country* (Cambridge: Cambridge University Press, 1985), pp. 4–13, 114–17; Christopher Lasch, *The True and Only Heaven: Progress and Its Critics* (New York: W. W. Norton, 1991).

14. Amnon Dankner and David Tartakover, *Where We Were and What We Did: An Israeli Lexicon of the Fifties and the Sixties* (in Hebrew) (Jerusalem: Keter, 1996).

15. Davis, *Yearning for Yesterday*, pp. 56–71; Ira Silver, "Role Transitions, Objects, and Identity," *Symbolic Interaction* 19, no. 1 (1996): 1–20.

16. Davis, *Yearning for Yesterday*, pp. 104–8. See also pp. 101–4.

17. Bernard Lewis, *History: Remembered, Recovered, Invented* (Princeton, N.J.: Princeton University Press, 1975), pp. 71–76.

18. See Georg Simmel, "The Persistence of Social Groups," *American Journal of Sociology* 3 (1897–98): 662–98.

19. David Hume, *A Treatise of Human Nature* (London: J. M. Dent, 1977), bk. 1, pt. 4, sec. 6, p. 240.

20. See also Sigmund Freud, *Civilization and Its Discontents* (New York: W. W. Norton, 1962), pp. 15–19.

21. Hume, *A Treatise of Human Nature*, bk. 1, pt. 4, sec. 6, pp. 245, 248.

22. Ibid., pp. 246–47.

23. Georg Simmel, "Bridge and Door," *Theory, Culture & Society* 11 (1994): 5–10.

24. Eviatar Zerubavel, *Patterns of Time in Hospital Life: A Sociological Perspective* (Chicago: University of Chicago Press, 1979), pp. 39–40, 136.

25. Kubler, *The Shape of Time*, p. 47.

26. Béla Balázs, *Theory of the Film: Character and Growth of a New Art* (New York: Dover, 1970), pp. 118ff.

27. Maurice Halbwachs, *The Collective Memory* (New York: Harper Colophon, 1980), pp. 128–57; Melinda J. Milligan, "Interactional Past and Potential: The Social Construction of Place Attachment," *Symbolic Interaction* 21 (1998): 8–15.

28. Kevin Lynch, *What Time Is This Place?* (Cambridge, Mass.: MIT Press, 1972), pp. 29–37; E. R. Chamberlin, *Preserving the Past* (London: J. M. Dent, 1979), pp. 51–64; Lowenthal, *The Past Is a Foreign Country*, pp. 275–78, 384–406.

29. Lowenthal, *The Past Is a Foreign Country*, p. 399.

30. Samuel C. Heilman, *A Walker in Jerusalem* (New York: Summit Books, 1986), pp. 77–111.

31. Ibid., pp. 80, 85, 89.

32. See, for example, Stefania Perring and Dominic Perring, *Then and Now* (New York: Macmillan, 1991); Giuseppe Gangi, *Rome Then and Now* (Rome: G & G Editrice, n.d.).

33. Shils, *Tradition*, pp. 69–71; Neil A. Silberman, *Between Past and Present: Archaeology, Ideology, and Nationalism in the Modern Middle East* (New York: Henry Holt, 1989).

34. *New York Times*, 12 December 1999, International section, p. 28.

35. Chamberlin, *Preserving the Past*, pp. 11–18; Silberman, *Between Past and Present*, pp. 87–101; Yael Zerubavel, *Recovered Roots: Collective Memory and the Making of Israeli National Tradition* (Chicago: University of Chicago Press, 1995), pp. 64–68, 129–33; Nachman Ben-Yehuda, *The Masada Myth: Collective Memory and Mythmaking in Israel* (Madison: University of Wisconsin Press, 1995). See also Y. Zerubavel, *Recovered Roots*, pp. 56–59, 185–89.

36. Y. Zerubavel, *Recovered Roots*, pp. 130–31. See also pp. 127–29, 135.

37. B. Lewis, *History*, pp. 101–2; Chamberlin, *Preserving the Past*, pp. 18–27.

38. Israel Gershoni and James P. Jankowski, *Egypt, Islam, and the Arabs: The Search for Egyptian Nationhood, 1900–1930* (New York: Oxford University Press, 1986), p. 147. Emphasis added.

39. Y. Zerubavel, *Recovered Roots*, pp. 120–21; Yael Zerubavel, "The Forest as a National Icon: Literature, Politics, and the Archeology of Memory," *Israel Studies* 1 (1996): 60–99.

40. Daniel Ben-Simon, "A Secure Step in a Sealed City" (in Hebrew), *Ha'aretz*, 28 August 1998, p. 14. Emphasis added.

41. D. W. Winnicott, "Transitional Objects and Transitional Phenomena," in *Playing and Reality* (London: Tavistock, 1971), pp. 1–25.

42. Silver, "Role Transitions, Objects, and Identity."

43. See also Kai T. Erikson, *Everything in Its Path: Destruction of Community in the Buffalo Creek Flood* (New York: Simon and Schuster, 1976), pp. 174–77.

44. Norimitsu Onishi, "A Tale of the Mullah and Muhammad's Amazing Cloak," *New York Times*, 19 December 2001, sec. B, p. 3.

45. Ian Wilson, *The Shroud of Turin: The Burial Cloth of Jesus Christ?* (Garden City, N.Y.: Doubleday, 1978).

46. Eric Davis, "The Museum and the Politics of Social Control in Modern Iraq," in *Commemorations: The Politics of National Identity*, edited by John R. Gillis (Princeton, N.J.: Princeton University Press, 1994), pp. 90–104; Tamar Katriel, *Performing the Past: A Study of Israeli Settlement Museums* (Mahwah, N.J.: Lawrence Erlbaum Associates, 1997).

47. Paul Zielbauer, "Found in Clutter, a Relic of Lincoln's Death," *New York Times*, 5 July 2001, sec. A, p.1—sec. B, p. 5.

48. Shils, *Tradition*, pp. 71–74.

49. Eviatar Zerubavel, *Social Mindscapes: An Invitation to Cognitive Sociology* (Cambridge, Mass.: Harvard University Press, 1997), p. 94.

50. Charles S. Peirce, *Collected Papers of Charles Sanders Peirce* (Cambridge, Mass.: Harvard University Press, 1962), 2:157–60; E. Zerubavel, *Social Mindscapes*, pp. 70–71, 137.

51. John F. Burns, "New Babylon Is Stalled by a Modern Upheaval," *New York Times*, 11 October 1990, International section, p. A13.

52. See also Shils, *Tradition*, p. 79; Lowenthal, *The Past Is a Foreign Country*, pp. 309–19.

53. Stephen Kern, *The Culture of Time and Space 1880–1918* (Cambridge, Mass.: Harvard University Press, 1983), p. 39; Lowenthal, *The Past Is a Foreign Country*, pp. 257–58, 367–68.

54. Eric J. Hobsbawm, "Introduction: Inventing Traditions," in *The Invention of Tradition*, edited by Eric J. Hobsbawm and Terence Ranger (Cambridge: Cambridge University Press, 1983), pp. 1–14.

55. Helene Henderson and Sue Ellen Thompson, eds., *Holidays, Festivals, and Celebrations of the World Dictionary*, 2d ed. (Detroit: Omnigraphics Inc., 1997), p. 230.

56. Hugh Trevor-Roper, "The Invention of Tradition: The Highland Tradition of Scotland," in *The Invention of Tradition*, edited by Eric J. Hobsbawm and Terence Ranger (Cambridge: Cambridge University Press, 1983), pp. 15–41.

57. See also Kubler, *The Shape of Time*, pp. 73–74; Hobsbawm, "Introduction: Inventing Traditions," p. 4; Paul Connerton, *How Societies Remember* (Cambridge: Cambridge University Press, 1989), pp. 45, 65–67.

58. Rick Bragg, "Emotional March Gains a Repentant Wallace," *New York Times*, 11 March 1995, sec. A, pp. 1, 9.

59. Karen A. Cerulo and Janet M. Ruane, "Death Comes Alive: Technology and the Re-

conception of Death," *Science as Culture* 6 (1997): 453–58. See also Alfred Schutz, "Making Music Together: A Study in Social Relationship," in *Collected Papers, vol. 2: Studies in Social Theory* (The Hague: Martinus Nijhoff, 1964), pp. 172–75.

60. Pam Belluck, "Pilgrims Wear Different Hats in Recast Thanksgiving Tales," *New York Times*, 23 November 1995, sec. A, p. 1; sec. B, p. 7.

61. Thomas R. Forrest, "Disaster Anniversary: A Social Reconstruction of Time," *Sociological Inquiry* 63 (1993): 444–56; Lyn Spillman, *Nation and Commemoration: Creating National Identities in the United States and Australia* (Cambridge: Cambridge University Press, 1997); Vered Vinitzky-Seroussi, *After Pomp and Circumstance: High School Reunion as an Autobiographical Occasion* (Chicago: University of Chicago Press, 1998).

62. Connerton, *How Societies Remember*, pp. 42–43.

63. Vered Vinitzky-Seroussi, "Commemorating a Difficult Past: Yitzhak Rabin's Memorials," *American Sociological Review* 67 (2002): 30–51. See also Y. Zerubavel, *Recovered Roots*, pp. 143–44.

64. Eviatar Zerubavel, "Calendars and History: A Comparative Study of the Social Organization of National Memory," in *States of Memory: Conflicts, Continuities, and Transformations in National Commemoration*, edited by Jeffrey K. Olick (Durham, N.C.: Duke University Press, in press). See also Mircea Eliade, *The Sacred and the Profane: The Nature of Religion* (New York: Harcourt, Brace & World, 1959), pp. 68–113.

65. See also Y. Zerubavel, *Recovered Roots*, p. 217.

66. Eviatar Zerubavel, *The Seven-Day Circle: The History and Meaning of the Week* (New York: Free Press, 1985), p. 84.

67. See also W. Lloyd Warner, *The Family of God* (New Haven, Conn.: Yale University Press, 1961), pp. 345–62.

68. E. Zerubavel, *The Seven-Day Circle*, pp. 51–54, 56–58.

69. See also E. Zerubavel, *Patterns of Time in Hospital Life*, pp. 6–8.

70. Ernest R. May, *"Lessons" of the Past: The Use and Misuse of History in American Foreign Policy* (New York: Oxford University Press, 1973); Richard E. Neustadt and Ernest R. May, *Thinking in Time: The Uses of History for Decision-Makers* (New York: Free Press, 1986); Yuen F. Khong, *Analogies at War: Korea, Munich, Dien Bien Phu, and the Vietnam Decisions of 1965* (Princeton, N.J.: Princeton University Press, 1992); Keith J. Holyoak and Paul Thagard, *Mental Leaps: Analogy in Creative Thought* (Cambridge, Mass.: MIT Press, 1995), pp. 101–9, 155–65; David B. Pillemer, *Momentous Events, Vivid Memories* (Cambridge, Mass.: Harvard University Press, 1998), pp. 79–83. See also Peter L. Berger and Thomas Luckmann, *The Social Construction of Reality: A Treatise in the Sociology of Knowledge* (Garden City, N.Y.: Doubleday, 1966), pp. 53–54; Samuel C. Heilman, *The People of the Book: Drama, Fellowship, and Religion* (Chicago: University of Chicago Press, 1983), p. 187; Yael Zerubavel, "The Death of Memory and the Memory of Death: Masada and the Holocaust as Historical Metaphors," *Representations* 45 (winter 1994): 72–100; Y. Zerubavel, *Recovered Roots*, pp. 160–67.

71. Warren Hoge, "Queen Breaks the Ice: Camilla's out of the Fridge," *New York Times*, 5 June 2000, sec. A, p. 4.

72. Katherine Stovel, "The Malleability of Precedent" (paper presented at the Annual Meeting of the Social Science History Association, New Orleans, 1996).

73. May, *"Lessons" of the Past*, p. 28.

74. Daniel Ben-Simon, "The Settlers' Nightmares" (in Hebrew), *Ha'aretz*, 23 June 2000, p. 16.

75. May, *"Lessons" of the Past*, pp. 97–99, 116; Khong, *Analogies at War*, pp. 59–61, 88–93.

76. Susan Sachs, "Bin Laden Images Mesmerize Muslims," *New York Times*, 9 October 2001, sec. B, p. 6.

77. Michael Walzer, *Exodus and Revolution* (New York: Basic Books, 1984), p. 39; Robert P. Hay, "George Washington: American Moses," *American Quarterly* 21 (1969): 780–91.

78. Robert Jervis, *Perception and Misperception in International Politics* (Princeton, N.J.: Princeton University Press, 1976), pp. 217, 221.

79. Pillemer, *Momentous Events, Vivid Memories*, p. 82.

80. Neustadt and May, *Thinking in Time*, pp. 48–53.

81. Zbigniew Brzezinski, "Can Communism Compete with the Olympics?" *New York Times*, 14 July 2001, sec. A, p. 15. See also the letter to the editor in the *New York Times*, 13 July 2001, sec. A, p. 20.

82. May, *"Lessons" of the Past*, pp. 99–100.

83. Ibid., pp. 6–18.

84. John McCain, interview by Bob Edwards, *Morning News*, National Public Radio, 14 September 2001.

85. Khong, *Analogies at War*, p. 6.

86. Robin Wagner-Pacifici, *Theorizing the Standoff: Contingency in Action* (Cambridge: Cambridge University Press, 2000), p. 93.

87. May, *"Lessons" of the Past*, p. 36.

88. Ibid., pp. 32, 52–53, 80–86; Neustadt and May, *Thinking in Time*, p. 36; Holyoak and Thagard, *Mental Leaps*, pp. 106, 156.

89. May, *"Lessons" of the Past*, p. 113; Khong, *Analogies at War*, pp. 59–61, 174–205.

90. Khong, *Analogies at War*, p. 5.

91. Ibid., p. 260; Howard Schuman and Cheryl Rieger, "Historical Analogies, Generational Effects, and Attitudes toward War," *American Sociological Review* 57 (1992): 315–26; Holyoak and Thagard, *Mental Leaps*, pp. 102–9.

92. Reinhart Koselleck, "Modernity and the Planes of Historicity," in *Futures Past: On the Semantics of Historical Time* (Cambridge, Mass.: MIT Press, 1985), p. 4.

93. Clifford Krauss, "Son of the Poor Is Elected in Peru over Ex-President," *New York Times*, 4 June 2001, sec. A, p. 1.

94. Gilmer W. Blackburn, *Education in the Third Reich: Race and History in Nazi Textbooks* (Albany: State University of New York Press, 1985), p. 54.

95. Y. Zerubavel, *Recovered Roots*, p. 73. See also pp. 70–76.

96. John Kifner, "Israeli and Palestinian Leaders Vow to Keep Working for Peace," *New York Times*, 27 July 2000, sec. A, p. 1; Evan Thomas, "The Road to September 11," *Newsweek*, 1 October 2001, p. 42. See also B. Lewis, *History*, pp. 83–87.

97. *Scott 1999 Standard Postage Stamp Catalogue* (Sidney, Ohio: Scott Publishing Co., 1998), 3:750. See also 3:747, 749; Nancy Cooper and Christopher Dickey, "After the War: Iraq's Designs," *Newsweek*, 8 August 1988, p. 35.

98. Wagner-Pacifici, *Theorizing the Standoff*, p. 93.

99. James Bennet, "Sharon Invokes Munich in Warning U.S. on 'Appeasement,'" *New York Times*, 5 October 2001, sec. A, p. 6.

100. Khong, *Analogies at War,* p. 137.

101. Ibid., p. 172. See also pp. 148–73.

102. For a long list of such "Purims," see "Special Purim," in *Encyclopaedia Judaica* (Jerusalem: Keter, 1972), 13:1396–1400. See also Henderson and Thompson, *Holidays, Festivals, and Celebrations of the World Dictionary,* p. 343.

103. Hay, "George Washington: American Moses," p. 782.

104. Khong, *Analogies at War,* pp. 5, 87.

105. Ibid., pp. 4, 258. See also pp. 76, 96, 259–60; R. W. Apple Jr., "A Military Quagmire Remembered: Afghanistan as Vietnam," *New York Times,* 31 October 2001, sec. B, pp. 1, 3.

106. Anthony DePalma, "In the War Cry of the Indians, Zapata Rides Again," *New York Times,* 27 January 1994, International section.

107. Nicholas D. Kristof, "With Genghis Revived, What Will Mongols Do?" *New York Times,* 23 March 1990, International section, p. A4.

108. Donald J. Wilcox, *The Measure of Times Past: Pre-Newtonian Chronologies and the Rhetoric of Relative Time* (Chicago: University of Chicago Press, 1987), p. 106.

109. See also ibid., pp. 52–53, 71–82, 123–25.

110. Edward E. Cummings, *Complete Poems* (New York: Harcourt Brace Jovanovich, 1972); James Joyce, *Ulysses* (New York: Random House, 1986).

111. Eviatar Zerubavel, *The Fine Line: Making Distinctions in Everyday Life* (New York: Free Press, 1991), p. 70.

112. Gershoni and Jankowski, *Egypt, Islam, and the Arabs,* pp. 143–63.

113. Vinitzky-Seroussi, *After Pomp and Circumstance,* pp. 113–31; Robert Zussman, "Autobiographical Occasions: Photography and the Representation of the Self" (paper presented at the Annual Meeting of the American Sociological Association, Chicago, August 1999).

114. See Erving Goffman, *Stigma: Notes on the Management of Spoiled Identity* (Englewood Cliffs, N.J.: Prentice-Hall, 1963), pp. 62–104.

115. See also ibid., pp. 75–76; Eviatar Zerubavel, "Personal Information and Social Life," *Symbolic Interaction* 5, no. 1 (1982): 104–5.

116. See Denise Grady, "Exchanging Obesity's Risks for Surgery's," *New York Times,* 12 October 2000, sec. A, pp. 1, 26.

117. See also Murray S. Davis, *Smut: Erotic Reality/Obscene Ideology* (Chicago: University of Chicago Press, 1983), pp. 107–8, 122–24; Jamie Mullaney, "Making It 'Count': Mental Weighing and Identity Attribution," *Symbolic Interaction* 22 (1999): 269–83.

118. Haskell Fain, *Between Philosophy and History: The Resurrection of Speculative Philosophy of History within the Analytic Tradition* (Princeton, N.J.: Princeton University Press, 1970), pp. 76–79.

119. See also Andrea Hood, "Editing the Life Course: Autobiographical Narratives, Identity Transformations, and Retrospective Framing" (unpublished manuscript, Rutgers University, Department of Sociology, 2002).

120. Harold Garfinkel, "Passing and the Managed Achievement of Sex Status in an 'Intersexed' Person," in *Studies in Ethnomethodology* (Englewood Cliffs, N.J.: Prentice-Hall, 1967), pp. 116–85.

121. See also Y. Zerubavel, *Recovered Roots,* pp. 15–36. On temporal "bracketing," see also Erving Goffman, *Frame Analysis: An Essay on the Organization of Experience* (New York: Harper Colophon, 1974), pp. 251–69.

Chapter Three

1. Yael Zerubavel, "Travels in Time and Space: Legendary Literature as a Vehicle for Shaping Collective Memory" (in Hebrew), *Teorya Uviqoret* 10 (summer 1997): 71–79. See also Yael Zerubavel, *Recovered Roots: Collective Memory and the Making of Israeli National Tradition* (Chicago: University of Chicago Press, 1995), pp. 92, 108.

2. Kenneth McNeil and James D. Thompson, "The Regeneration of Social Organizations," *American Sociological Review* 36 (1971): 624–37.

3. See also Samuel C. Heilman, *The People of the Book: Drama, Fellowship, and Religion* (Chicago: University of Chicago Press, 1983), p. 62.

4. See also Alfred Schutz and Thomas Luckmann, *The Structures of the Life-World* (Evanston, Ill.: Northwestern University Press, 1973), pp. 87–92.

5. See also Raymond L. Schmitt, "Symbolic Immortality in Ordinary Contexts: Impediments to the Nuclear Era," *Omega* 13 (1982–83): 95–116.

6. Bill Rolston, *Drawing Support: Murals in the North of Ireland* (Belfast: Beyond the Pale Publications, 1992).

7. See, for example, Ferdinand Tönnies, *Community and Society* (New York: Harper Torchbooks, 1963), pp. 42–43.

8. Alex Shoumatoff, *The Mountain of Names: A History of the Human Family* (New York: Simon and Schuster, 1985), p. 217.

9. Hugh Baker, *Chinese Family and Kinship* (New York: Columbia University Press, 1979), p. 26.

10. Israel Gershoni and James P. Jankowski, *Egypt, Islam, and the Arabs: The Search for Egyptian Nationhood, 1900–1930* (New York: Oxford University Press, 1986), pp. 165–66; Stephen Howe, *Afrocentrism: Mythical Pasts and Imagined Homes* (London: Verso, 1998), pp. 37, 43.

11. Shoumatoff, *The Mountain of Names*, p. 55. See also Randall Collins, *The Sociology of Philosophies: A Global Theory of Intellectual Change* (Cambridge, Mass.: Harvard University Press, 1998), pp. 54–58, 64–68.

12. Shoumatoff, *The Mountain of Names*, p. 89.

13. Ibid., p. 72.

14. Frederick Allen, "They're Still There: The Oldest Business in America," *American Heritage of Invention and Technology* 15, no. 3 (2000): 6.

15. Max Weber, *Economy and Society: An Outline of Interpretive Sociology* (Berkeley and Los Angeles: University of California Press, 1978), pp. 1139–41. See also pp. 246–48, 1121–25, 1135–39.

16. See Baker, *Chinese Family and Kinship*, pp. 26–27.

17. Guy Murchie, *The Seven Mysteries of Life: An Exploration in Science and Philosophy* (New York: Mariner Books, 1999 [1978]), p. 357 (page citations are to the reprint edition); Richard Dawkins, *River out of Eden: A Darwinian View of Life* (New York: Basic Books, 1995).

18. On the latter, see Stanley Milgram, "The Small World Problem," in *The Individual in a Social World: Essays and Experiments,* 2d ed. (New York: McGraw-Hill, 1992 [1967]), pp. 259–75; Ithiel de Sola Pool and Manfred Kochen, "Contacts and Influence," in *The Small World,* edited by Manfred Kochen (Norwood, N.J.: Ablex, 1989), pp. 3–51.

19. See also Ruth Simpson, "I Was There: Establishing Ownership of Historical Moments" (paper presented at the Annual Meeting of the American Sociological Association, Los Angeles, 1994).

20. Matthew L. Chancey, "Mrs. Alberta Martin: The Old Man's Darling," <http://lastconfederatewidow.com>, accessed 7 February 2002.

21. Patricia Polacco, *Pink and Say* (New York: Philomel Books, 1994).

22. See, for example, Richard Lewontin, *Human Diversity* (New York: Scientific American Books, 1982), p. 162; Donald Johanson and Blake Edgar, *From Lucy to Language* (New York: Simon and Schuster, 1996), p. 112; Collins, *The Sociology of Philosophies*, pp. 54–79.

23. John Guare, *Six Degrees of Separation* (New York: Random House, 1990).

24. Milgram, "The Small World Problem."

25. Avraham S. Friedberg, *Zikhronot le-Veit David* (in Hebrew) (Ramat Gan, Israel: Masada, 1958).

26. Kenneth W. Wachter, "Ancestors at the Norman Conquest," in *Genealogical Demography*, edited by Bennett Dyke and Warren T. Morrill (New York: Academic Press, 1980), p. 92; Edward Shils, *Tradition* (Chicago: University of Chicago Press, 1981), p. 37.

27. See also Shoumatoff, *The Mountain of Names*, p. 73.

28. Georg Simmel, "The Persistence of Social Groups," *American Journal of Sociology* 3 (1897–98): 669–71. See also David Hume, *A Treatise of Human Nature* (London: J. M. Dent, 1977), bk. 1, pt. 4, sec. 6, pp. 243–44.

29. Simmel, "The Persistence of Social Groups," p. 669.

30. Karl Mannheim, "The Problem of Generations," in *Essays on the Sociology of Knowledge* (London: Routledge and Kegan Paul, 1951), pp. 292–94.

31. Simmel, "The Persistence of Social Groups," p. 670. See also Hume, *A Treatise of Human Nature*, bk. 1, pt. 4, sec. 6, p. 242.

32. See also Eviatar Zerubavel, *Patterns of Time in Hospital Life: A Sociological Perspective* (Chicago: University of Chicago Press, 1979), pp. 46–50, 60–61.

33. See also Simmel, "The Persistence of Social Groups," pp. 671–75; E. Zerubavel, *Patterns of Time in Hospital Life*, pp. 43–46.

34. On the inverse relation between gaps and continuity, see Eviatar Zerubavel, *The Fine Line: Making Distinctions in Everyday Life* (New York: Free Press, 1991), pp. 21–32.

35. See also Haskell Fain, *Between Philosophy and History: The Resurrection of Speculative Philosophy of History within the Analytic Tradition* (Princeton, N.J.: Princeton University Press, 1970), p. 78.

36. Marshall D. Johnson, *The Purpose of the Biblical Genealogies with Special Reference to the Setting of the Genealogies of Jesus* (London: Cambridge University Press, 1969), p. 78.

37. Shoumatoff, *The Mountain of Names*, pp. 67, 72. See also pp. 66, 71.

38. Thomas A. Hale, *Griots and Griottes: Masters of Words and Music* (Bloomington: Indiana University Press, 1998), p. 124.

39. Seth Faison, "Not Equal to Confucius, but Friends to His Memory," *New York Times*, 10 October 1997, International section.

40. Johnson, *The Purpose of the Biblical Genealogies*, p. 79.

41. Mary Bouquet, "Family Trees and Their Affinities: The Visual Imperative of the Genealogical Diagram," *Journal of the Royal Anthropological Institute*, n.s., 2 (1996): 47.

42. Shoumatoff, *The Mountain of Names*, p. 64; Jessica Libove, "Guardians of Collective Memory: The Mnemonic Functions of the Griot in West Africa" (unpublished manuscript, Rutgers University, Department of Anthropology, 2000). See also Anthony

Wagner, "Bridges to Antiquity," in *Pedigree and Progress: Essays in the Genealogical Interpretation of History* (London: Phillimore, 1975), pp. 50–75.

43. E. Zerubavel, *The Fine Line*, pp. 56–57.

44. Virginia R. Domínguez, *White by Definition: Social Classification in Creole Louisiana* (New Brunswick, N.J.: Rutgers University Press, 1986), pp. 188–204; France W. Twine, *Racism in a Racial Democracy: The Maintenance of White Supremacy in Brazil* (New Brunswick, N.J.: Rutgers University Press, 1998), pp. 116–33.

45. Frank Bruni and Katharine Q. Seelye, "Campaign Contrasts Grow Starker," *New York Times,* 2 July 2000, sec. A, p. 11.

46. Carey Goldberg, "DNA Offers Link to Black History," *New York Times,* 28 August 2000, sec. A, p. 10.

47. Baker, *Chinese Family and Kinship,* p. 95.

48. Shoumatoff, *The Mountain of Names,* p. 50.

49. Lewis H. Morgan, *Systems of Consanguinity and Affinity of the Human Family* (Lincoln: University of Nebraska Press, 1997), p. 17.

50. Tönnies, *Community and Society,* pp. 42, 48.

51. Morgan, *Systems of Consanguinity and Affinity,* p. 10.

52. J. D. Freeman, "On the Concept of the Kindred," in *Kinship and Social Organization,* edited by Paul Bohannan and John Middleton (Garden City, N.Y.: American Museum of Natural History, 1968), p. 255; Meyer Fortes, "Descent, Filiation, and Affinity," in *Time and Social Structure and Other Essays* (London: Athlone Press, 1970), p. 111; Alfred R. Radcliffe-Brown, "The Study of Kinship Systems," in *Structure and Function in Primitive Society* (New York: Free Press, 1965), pp. 51–53.

53. Millicent R. Ayoub, "The Family Reunion," *Ethnology* 5 (1966): 416, 418.

54. Baker, *Chinese Family and Kinship,* pp. 90–91. See also Ernest L. Schusky, *Variation in Kinship* (New York: Holt, Rinehart, and Winston, 1974), p. 53.

55. Morgan, *Systems of Consanguinity and Affinity,* pp. 10–11, 25. See also David M. Schneider, *American Kinship: A Cultural Account* (Englewood Cliffs, N.J.: Prentice-Hall, 1968), pp. 25, 65.

56. Theodore D. McCown and Kenneth A. R. Kennedy, eds., *Climbing Man's Family Tree: A Collection of Major Writings on Human Phylogeny, 1699 to 1971* (Englewood Cliffs, N.J.: Prentice-Hall, 1972), p. 10. See also Bouquet, "Family Trees and Their Affinities," p. 63.

57. See also Edward E. Evans-Pritchard, *The Nuer: A Description of the Modes of Livelihood and Political Institutions of a Nilotic People* (London: Oxford University Press, 1940), pp. 106–8.

58. Morris Swadesh, "What Is Glottochronology?" in *The Origin and Diversification of Language* (Chicago: Aldine, 1971), pp. 271–76; Colin Renfrew, *Archaeology and Language: The Puzzle of Indo-European Origins* (New York: Cambridge University Press, 1987), pp. 101, 113–15, 118.

59. For some etymological evidence, see, for example, Morgan, *Systems of Consanguinity and Affinity,* pp. 95, 106, 314, 349, 552, 555.

60. Dawkins, *River out of Eden,* p. 35.

61. See also Wachter, "Ancestors at the Norman Conquest," p. 92; Shoumatoff, *The Mountain of Names,* p. 245.

62. Schneider, *American Kinship,* pp. 67–68.

63. See also ibid., p. 73; Shoumatoff, *The Mountain of Names*, p. 22.

64. Evans-Pritchard, *The Nuer*, pp. 106–7, 200–201; Meyer Fortes, "The Significance of Descent in Tale Social Structure," in *Time and Social Structure and Other Essays* (London: Athlone Press, 1970), p. 37.

65. Baker, *Chinese Family and Kinship*, p. 104.

66. See, for example, Gershoni and Jankowski, *Egypt, Islam, and the Arabs*, p. 165.

67. Anthony D. Smith, *The Ethnic Origins of Nations* (Oxford: Basil Blackwell, 1986), pp. 24–26.

68. James Bennet, "Hillary Clinton, in Morocco, Says NATO Attack Aims at Stopping Bloodshed," *New York Times*, 31 March 1999, International section, p. A10.

69. Mirta Ojito, "Blacks on a Brooklyn Street: Both Cynics and Optimists Speak Out," *New York Times*, 26 March 1998, International section, p. A13.

70. Schusky, *Variation in Kinship*, pp. 53–54; Baker, *Chinese Family and Kinship*, pp. 107–11.

71. Freeman, "On the Concept of the Kindred," pp. 261, 265.

72. See also Northcote W. Thomas, *Kinship Organisations and Group Marriage in Australia* (New York: Humanities Press, 1966), pp. 3–4.

73. Alfred R. Radcliffe-Brown, "Patrilineal and Matrilineal Succession," in *Structure and Function in Primitive Society* (New York: Free Press, 1965), pp. 32–48.

74. Shoumatoff, *The Mountain of Names*, p. 31; Nancy Jay, *Throughout Your Generations Forever: Sacrifice, Religion, and Paternity* (Chicago: University of Chicago Press, 1992), p. 30.

75. Shoumatoff, *The Mountain of Names*, p. 37. See also Freeman, "On the Concept of the Kindred," p. 262.

76. Jay, *Throughout Your Generations Forever*, p. 47. See also John R. Gillis, *A World of Their Own Making: Myth, Ritual, and the Quest for Family Values* (New York: Basic Books, 1996), p. 184; Katherine Verdery, *The Political Lives of Dead Bodies: Reburial and Postsocialist Change* (New York: Columbia University Press, 1999), p. 118.

77. Radcliffe-Brown, "Patrilineal and Matrilineal Succession," p. 47.

78. See also Julie M. Gricar, "How Thick Is Blood? The Social Construction and Cultural Configuration of Kinship" (Ph.D. diss., Columbia University, 1991), p. 323; Johanna E. Foster, "Feminist Theory and the Politics of Ambiguity: A Comparative Analysis of the Multiracial Movement, the Intersex Movement and the Disability Rights Movement as Contemporary Struggles over Social Classification in the United States" (Ph.D. diss., Rutgers University, 2000), pp. 73–74.

79. Talcott Parsons, "The Kinship System of the Contemporary United States," in *Essays in Sociological Theory*, rev. ed. (New York: Free Press, 1964 [1943]), p. 184; Raymond Firth, "A Note on Descent Groups in Polynesia," in *Kinship and Social Organization*, edited by Paul Bohannan and John Middleton (Garden City, N.Y.: American Museum of Natural History, 1968 [1957]), p. 219; Freeman, "On the Concept of the Kindred," p. 271; Edmund Leach, "On Certain Unconsidered Aspects of Double Descent Systems," *Man* 62 (1962): 132; Schusky, *Variation in Kinship*, pp. 26–39; Shoumatoff, *The Mountain of Names*, p. 34. See also Ayoub, "The Family Reunion," p. 431.

80. See also Shoumatoff, *The Mountain of Names*, p. 244.

81. Murchie, *The Seven Mysteries of Life*, p. 351.

82. William W. Howells, "The Dispersion of Modern Humans," in *The Cambridge En-*

cyclopedia of Human Evolution, edited by Steve Jones et al. (Cambridge: Cambridge University Press, 1992), p. 400; Luigi L. Cavalli-Sforza and Francesco Cavalli-Sforza, *The Great Human Diasporas: The History of Diversity and Evolution* (Reading, Mass.: Addison-Wesley, 1995), pp. 121–23; Luigi L. Cavalli-Sforza, Paolo Menozzi, and Alberto Piazza, *The History and Geography of Human Genes,* abridged pbk. ed. (Princeton, N.J.: Princeton University Press, 1996), p. 94; Nicholas Wade, "To People the World, Start With 500," *New York Times,* 11 November 1997, sec. F, p. 3; Luigi L. Cavalli-Sforza, *Genes, Peoples, and Languages* (New York: North Point Press, 2000), pp. 60–63; Nicholas Wade, "The Human Family Tree: 10 Adams and 18 Eves," *New York Times,* 2 May 2000, sec. F, pp. 1–5; Nicholas Wade, "The Origin of the Europeans," *New York Times,* 14 November 2000, sec. F, pp. 1–9.

83. See also Lewontin, *Human Diversity,* pp. 161–62.

84. Dawkins, *River out of Eden,* p. 52.

85. See Alan R. Rogers and Lynn B. Jorde, "Genetic Evidence on Modern Human Origins," *Human Biology* 67 (1995): 21–22; Christopher Stringer and Robin McKie, *African Exodus: The Origins of Modern Humanity* (New York: Henry Holt, 1997), pp. 116, 182.

86. Rebecca L. Cann, Mark Stoneking, and Allan C. Wilson, "Mitochondrial DNA and Human Evolution," *Nature* 325 (1987): 31–36; Maryellen Ruvolo et al., "Mitochondrial COII Sequences and Modern Human Origins," *Molecular Biology and Evolution* 10 (1993): 1115–35; Rogers and Jorde, "Genetic Evidence on Modern Human Origins," p. 25; Johanson and Edgar, *From Lucy to Language,* p. 56; Natalie Angier, "Do Races Differ? Not Really, Genes Show," *New York Times,* 22 August 2000, sec. F, p. 6.

87. Cavalli-Sforza and Cavalli-Sforza, *The Great Human Diasporas,* pp. 114–16, 123–25; Stringer and McKie, *African Exodus,* p. 162.

88. Angier, "Do Races Differ?", sec. F, p. 1.

89. Stringer and McKie, *African Exodus,* p. 117. See also Angier, "Do Races Differ?", sec. F, p. 1.

90. Stringer and McKie, *African Exodus,* p. 177.

91. George W. Stocking, "French Anthropology in 1800," in *Race, Culture, and Evolution: Essays in the History of Anthropology* (New York: Free Press, 1968 [1964]), p. 39; George W. Stocking, "The Persistence of Polygenist Thought in Post-Darwinian Anthropology," in *Race, Culture, and Evolution: Essays in the History of Anthropology* (New York: Free Press, 1968), pp. 44–45; McCown and Kennedy, *Climbing Man's Family Tree,* p. 32; Richard H. Popkin, "The Pre-Adamite Theory in the Renaissance," in *Philosophy and Humanism: Renaissance Essays in Honor of Paul Oskar Kristeller,* edited by Edward P. Mahoney (New York: Columbia University Press, 1976), pp. 57–58, 66–69; Peter J. Bowler, *The Invention of Progress: The Victorians and the Past* (Oxford: Basil Blackwell, 1989), p. 107; Clive Gamble, *Timewalkers: The Prehistory of Global Colonization* (Cambridge, Mass.: Harvard University Press, 1994), p. 25; Benjamin Braude, "The Sons of Noah and the Construction of Ethnic and Geographical Identities in the Medieval and Early Modern Periods," *The William and Mary Quarterly,* 3d ser., 54 (1997): 103–42.

92. Nancy Stepan, *The Idea of Race in Science: Great Britain 1800–1960* (Hamden, Conn.: Archon Books, 1982), p. 29; Peter J. Bowler, *Theories of Human Evolution: A Century of Debate, 1844–1944* (Baltimore: Johns Hopkins University Press, 1986), pp. 127, 140; Ian Tattersall and Jeffrey H. Schwartz, *Extinct Humans* (Boulder, Colo.: Westview,

2000), p. 20.

93. Bowler, *Theories of Human Evolution*, p. 127.

94. Stepan, *The Idea of Race in Science*, p. 29.

95. Howe, *Afrocentrism*, pp. 73–74, 227, 271.

96. Arthur de Gobineau, *The Inequality of Human Races* (New York: Howard Fertig, 1967), p. 137. See also pp. 139–40; Howe, *Afrocentrism*, pp. 73–74.

97. See also Stepan, *The Idea of Race in Science*, p. 106; Bowler, *Theories of Human Evolution*, pp. 127–28, 131; Bowler, *The Invention of Progress*, p. 120.

98. Carl Vogt, *Lectures on Man: His Place in Creation and in the History of the Earth* (London: Longman, Green, Longman, and Roberts, 1864), pp. 172, 214, 222, 401–4, 440, 465–67.

99. Hermann Klaatsch, *The Evolution and Progress of Mankind* (New York: Frederick A. Stokes, 1923), pp. 105–6, 269–84; Bowler, *Theories of Human Evolution*, pp. 135–37, 141. See also Paul Topinard, *Anthropology* (London: Chapman & Hall, 1878), pp. 510–11, 518; Stocking, "The Persistence of Polygenist Thought in Post-Darwinian Anthropology," pp. 57, 63, 68; Stepan, *The Idea of Race in Science*, pp. 106–8; Bowler, *The Invention of Progress*, p. 119.

100. Franz Weidenreich, "Facts and Speculations concerning the Origin of *Homo sapiens*," in *Climbing Man's Family Tree: A Collection of Major Writings on Human Phylogeny, 1699 to 1971*, edited by Theodore D. McCown and Kenneth A. R. Kennedy (Englewood Cliffs, N.J.: Prentice-Hall, 1972 [1947]), pp. 351–53 (page citations are to the reprint edition); Carleton S. Coon, *The Origin of Races* (New York: Alfred A. Knopf, 1962), pp. 335, 371–587. See also the figures in Howells, "The Dispersion of Modern Humans," p. 392; Göran Burenhult, "Modern People in Africa and Europe," in *The First Humans: Human Origins and History to 10,000 B.C.*, edited by Göran Burenhult (New York: HarperCollins, 1993), p. 80.

101. Howells, "The Dispersion of Modern Humans," p. 390; Burenhult, "Modern People in Africa and Europe," p. 80; Colin Groves, "Human Origins," in *The First Humans: Human Origins and History to 10,000 B.C.*, edited by Göran Burenhult (New York: HarperCollins, 1993), p. 49; Ian Tattersall, *The Fossil Trail: How We Know What We Think We Know about Human Evolution* (New York: Oxford University Press, 1995), p. 214. See also Bowler, *Theories of Human Evolution*, pp. 55–56, 127, 140, 188; Bowler, *The Invention of Progress*, p. 119.

102. Alan G. Thorne and Milford H. Wolpoff, "Regional Continuity in Australasian Pleistocene Hominid Evolution," *American Journal of Physical Anthropology* 55 (1981): 341–42. See also Milford H. Wolpoff et al., "Modern Human Origins," *Science* 241 (1988): 772–73.

103. Coon, *The Origin of Races*, p. 37. See also p. viii.

104. Thorne and Wolpoff, "Regional Continuity in Australasian Pleistocene Hominid Evolution," p. 337. See also Tattersall, *The Fossil Trail*, p. 216.

105. Lewontin, *Human Diversity*, p. 164.

106. John Gribbin, "Human vs. Gorilla: The 1% Advantage," *Science Digest* 90 (August 1982): 74. See also Jerold Lowenstein and Adrienne Zihlman, "The Invisible Ape," *New Scientist*, 3 December 1988, p. 57.

107. Bowler, *Theories of Human Evolution*, p. 245; Bernard A. Wood, "Evolution and Australopithecines," in *The Cambridge Encyclopedia of Human Evolution*, edited by Steve

Jones et al. (Cambridge: Cambridge University Press, 1992), p. 240; Groves, "Human Origins," pp. 50–51; Gamble, *Timewalkers*, p. 53; Tattersall and Schwartz, *Extinct Humans*, pp. 116, 244. See also John Reader, *Missing Links: The Hunt for Earliest Man* (Boston: Little, Brown, & Co., 1981), pp. 192–94, 212–13; Lewontin, *Human Diversity*, pp. 163–64.

108. Bowler, *Theories of Human Evolution*, p. 35; Stephen J. Gould, *Wonderful Life: The Burgess Shale and the Nature of History* (New York: W. W. Norton, 1989), p. 319; Groves, "Human Origins," pp. 49–51; Dawkins, *River out of Eden*, p. 53.

109. Michael Hammond, "The Expulsion of the Neanderthals from Human Ancestry: Marcellin Boule and the Social Context of Scientific Research," *Social Studies of Science* 12 (1982): 1–36; Bowler, *Theories of Human Evolution*, pp. 75–111. See also Gould, *Wonderful Life*, pp. 29–31; Bowler, *The Invention of Progress*, pp. 101–2, 121, 124–27; Tattersall and Schwartz, *Extinct Humans*, p. 244.

110. See Bowler, *The Invention of Progress*, p. 122.

111. Tattersall, *The Fossil Trail*, p. 4. See also Ramona Morris and Desmond Morris, *Men and Apes* (New York: Bantam, 1968), p. 145.

112. Tattersall and Schwartz, *Extinct Humans*, p. 22.

113. Tattersall, *The Fossil Trail*, p. 4; Harriet Ritvo, "Border Trouble: Shifting the Line between People and Other Animals," *Social Research* 62 (1995): 484; Tattersall and Schwartz, *Extinct Humans*, pp. 23–24.

114. Arthur O. Lovejoy, *The Great Chain of Being: A Study of the History of an Idea* (Cambridge, Mass.: Harvard University Press, 1936), p. 234.

115. Ibid., pp. 24–241. See also McCown and Kennedy, *Climbing Man's Family Tree*, p. 6; Stepan, *The Idea of Race in Science*, p. 13.

116. Lester Crocker, "Diderot and Eighteenth Century French Transformism," in *Forerunners of Darwin: 1745–1859*, edited by Bentley Glass et al. (Baltimore: Johns Hopkins University Press, 1959), pp. 129–31; McCown and Kennedy, *Climbing Man's Family Tree*, p. 12. See also Lovejoy, *The Great Chain of Being*, pp. 278–79.

117. Rulon Wells, "The Life and Growth of Language: Metaphors in Biology and Linguistics," in *Biological Metaphor and Cladistic Classification: An Interdisciplinary Perspective*, edited by Henry M. Hoenigswald and Linda F. Wiener (Philadelphia: University of Pennsylvania Press, 1987), p. 73.

118. Jean-Baptiste Lamarck, *Zoological Philosophy: An Exposition with Regard to the Natural History of Animals* (New York: Hafner, 1963), p. 39.

119. Ibid., p. 170. Emphasis added. See also pp. 172–73.

120. Owsei Temkin, "The Idea of Descent in Post-Romantic German Biology: 1848–1858," in *Forerunners of Darwin: 1745–1859*, edited by Bentley Glass et al. (Baltimore: Johns Hopkins University Press, 1959), pp. 342–51; Arthur O. Lovejoy, "The Argument for Organic Evolution before the Origin of Species, 1830–1858," in *Forerunners of Darwin: 1745–1859*, edited by Bentley Glass et al. (Baltimore: Johns Hopkins University Press, 1959), pp. 356–414; Tattersall, *The Fossil Trail*, p. 13.

121. Robert Chambers, *Vestiges of the Natural History of Creation* (Chicago: University of Chicago Press, 1994), pp. 219, 234.

122. Bowler, *The Invention of Progress*, pp. 89–91, 139–43; James A. Secord, introduction to *Vestiges of the Natural History of Creation*, by Robert Chambers (Chicago: University of Chicago Press, 1994), pp. ix–x.

123. Charles Darwin, *The Origin of Species* (New York: Mentor Books, 1958), pp. 391–92. See also p. 393; Charles Darwin, *The Descent of Man and Selection in Relation to Sex* (Amherst, N.Y.: Prometheus, 1998), pp. 153–55.

124. Darwin, *The Origin of Species*, pp. 391, 394–95. See also Robert J. Richards, *The Meaning of Evolution: The Morphological Construction and Ideological Reconstruction of Darwin's Theory* (Chicago: University of Chicago Press, 1992), p. 165; Tattersall, *The Fossil Trail*, p. 19; Peter J. Bowler, *Life's Splendid Drama: Evolutionary Biology and the Reconstruction of Life's Ancestry, 1860–1940* (Chicago: University of Chicago Press, 1996), pp. 41, 51; Tattersall and Schwartz, *Extinct Humans*, p. 44.

125. Darwin, *The Origin of Species*, pp. 114–22. See also Richards, *The Meaning of Evolution*, pp. 110–11.

126. Darwin, *The Origin of Species*, p. 391.

127. Bowler, *Theories of Human Evolution*, p. 2.

128. Ernst Haeckel, *The Evolution of Man: A Popular Exposition of the Principal Points of Human Ontogeny and Phylogeny* (New York: D. Appleton, 1879 [1874]), p. 102; Ernst Haeckel, *Anthropogenie oder Entwickelungsgeschichte des Menschen* (Leipzig: Wilhelm Engelmann, 1874), p. 496; Stephen J. Gould, *Ontogeny and Phylogeny* (Cambridge, Mass.: Harvard University Press, 1977), pp. 76–77; Jane M. Oppenheimer, "Haeckel's Variations on Darwin," in *Biological Metaphor and Cladistic Classification: An Interdisciplinary Perspective*, edited by Henry M. Hoenigswald and Linda F. Wiener (Philadelphia: University of Pennsylvania Press, 1987), p. 124; Bowler, *The Invention of Progress*, p. 155; Gould, *Wonderful Life*, p. 263; Bouquet, "Family Trees and Their Affinities," p. 57; Johanson and Edgar, *From Lucy to Language*, p. 37.

129. Reader, *Missing Links*, p. 40. See also Tattersall, *The Fossil Trail*, pp. 28–29.

130. Reader, *Missing Links*, pp. 41, 47–48, 50; Tattersall, *The Fossil Trail*, pp. 35–36; Johanson and Edgar, *From Lucy to Language*, p. 187.

131. Adrian Desmond, *Archetypes and Ancestors: Palaeontology in Victorian London 1850–1875* (Chicago: University of Chicago Press, 1984), pp. 156–57.

132. Darwin, *The Descent of Man*, p. 3.

133. Ibid., pp. 160, 630.

134. George H. Nuttall, *Blood Immunity and Blood Relationship: A Demonstration of Certain Blood-Relationships amongst Animals by means of the Precipitin Test for Blood* (Cambridge: Cambridge University Press, 1904), pp. 1–4, 319. See also Tattersall, *The Fossil Trail*, pp. 122–23.

135. Thomas H. Huxley, *Evidence as to Man's Place in Nature* (Ann Arbor: University of Michigan Press, 1959), p. 123. See also pp. 83–86.

136. Morris Goodman, "Serological Analysis of the Systematics of Recent Hominoids," *Human Biology* 35 (1963): 399–400; Morris Goodman, "Reconstructing Human Evolution from Proteins," in *The Cambridge Encyclopedia of Human Evolution*, edited by Steve Jones et al. (Cambridge: Cambridge University Press, 1992), pp. 307–8. See also Vincent Sarich and Allan C. Wilson, "Immunological Time Scale for Hominid Evolution," *Science* 158 (1967): 1200–1203; Lowenstein and Zihlman, "The Invisible Ape," pp. 56–57; Vincent Sarich, "Immunological Evidence on Primates," in *The Cambridge Encyclopedia of Human Evolution*, edited by Steve Jones et al. (Cambridge: Cambridge University Press, 1992), p. 306; Charles G. Sibley, "DNA-DNA Hybridisation in the Study of Primate Evolution," in *The Cambridge Encyclopedia of Human Evolution*, edited by Steve Jones et al. (Cambridge: Cambridge University Press, 1992),

p. 313; Johanson and Edgar, *From Lucy to Language*, pp. 30–32.

137. Goodman, "Reconstructing Human Evolution from Proteins," p. 310; Jared Diamond, *The Third Chimpanzee: The Evolution and Future of the Human Animal* (New York: HarperCollins, 1992), p. 23; Johanson and Edgar, *From Lucy to Language*, pp. 32, 111.

138. Stringer and McKie, *African Exodus*, p. 21. See also Gribbin, "Human vs. Gorilla," p. 73; Sarich, "Immunological Evidence on Primates," pp. 304–5.

139. Cavalli-Sforza and Cavalli-Sforza, *The Great Human Diasporas*, pp. 34–37; Tattersall, *The Fossil Trail*, pp. 124–25; Cavalli-Sforza, *Genes, Peoples, and Languages*, p. 78.

140. Lowenstein and Zihlman, "The Invisible Ape," p. 59.

141. Sarich and Wilson, "Immunological Time Scale for Hominid Evolution," p. 1202; Lowenstein and Zihlman, "The Invisible Ape," pp. 57–59; Goodman, "Reconstructing Human Evolution from Proteins," p. 308; Sarich, "Immunological Evidence on Primates," pp. 305–6; Jay Kelley, "Evolution of Apes," in *The Cambridge Encyclopedia of Human Evolution*, edited by Steve Jones et al. (Cambridge: Cambridge University Press, 1992), p. 230; Adrian E. Friday, "Human Evolution: The Evidence from DNA Sequencing," in *The Cambridge Encyclopedia of Human Evolution*, edited by Steve Jones et al. (Cambridge: Cambridge University Press, 1992), p. 320; Diamond, *The Third Chimpanzee*, p. 24; Groves, "Human Origins," p. 43; John N. Wilford, "A Fossil Unearthed in Africa Pushes Back Human Origins," *New York Times*, 11 July 2002, sec. A, pp. 1, 12.

142. Gribbin and Cherfas, *The Monkey Puzzle*, p. 29.

143. Ibid., p. 17. See also Johanson and Edgar, *From Lucy to Language*, p. 33.

144. See also Murchie, *The Seven Mysteries of Life*, pp. 357–62; Goodman, "Reconstructing Human Evolution from Proteins," p. 311; Dawkins, *River out of Eden*, pp. 8, 12.

145. See also Cavalli-Sforza and Cavalli-Sforza, *The Great Human Diasporas*, p. 38.

146. See Morgan, *Systems of Consanguinity and Affinity*, p. 25; Jack Goody, *The Development of the Family and Marriage in Europe* (Cambridge: Cambridge University Press, 1983), pp. 136–38.

147. See also E. Zerubavel, *The Fine Line*, pp. 61–80.

148. See also George G. Simpson, "The Meaning of Taxonomic Statements," in *Naming Our Ancestors: An Anthology of Hominid Taxonomy*, edited by W. Eric Meikle and Sue T. Parker (Prospect Heights, Ill.: Waveland, 1994 [1963]), pp. 179–81; Colin P. Groves and Vratislav Mazák, "An Approach to the Taxonomy of the Hominidae: Gracile Villafranchian Hominids in Africa," in *Naming Our Ancestors: An Anthology of Hominid Taxonomy*, edited by W. Eric Meikle and Sue T. Parker (Prospect Heights, Ill.: Waveland Press, 1994 [1975]), pp. 108–9; Robert Martin, "Classification and Evolutionary Relationships," in *The Cambridge Encyclopedia of Human Evolution*, edited by Steve Jones et al. (Cambridge: Cambridge University Press, 1992), pp. 17–18; W. Eric Meikle and Sue T. Parker, "Introduction: Names, Binomina, and Nomenclature in Paleoanthropology," in *Naming Our Ancestors: An Anthology of Hominid Taxonomy* (Prospect Heights, Ill.: Waveland Press, 1994), p. 9.

149. Tattersall, *The Fossil Trail*, p. 75.

150. Ibid.

151. See, for example, Gamble, *Timewalkers*, pp. 148–76; Tattersall, *The Fossil Trail*, pp. 114–15; John N. Wilford, "When Humans Became Human," *New York Times*, 26 February 2002, sec. F, pp. 1, 5.

152. See, for example, Christopher Stringer, "Evolution of Early Humans," in *The Cambridge Encyclopedia of Human Evolution*, edited by Steve Jones et al. (Cambridge: Cambridge University Press, 1992), pp. 245–46; Gamble, *Timewalkers*, p. 135; Tattersall, *The Fossil Trail*, pp. 173–75.

153. Reader, *Missing Links*, p. 189.

154. Ernst Mayr, "The Taxonomic Evaluation of Fossil Hominids," in *Climbing Man's Family Tree: A Collection of Major Writings on Human Phylogeny, 1699 to 1971*, edited by Theodore D. McCown and Kenneth A. R. Kennedy (Englewood Cliffs, N.J.: Prentice-Hall, 1972), p. 380. See also p. 381; G. Simpson, "The Meaning of Taxonomic Statements," p. 180; Martin, "Classification and Evolutionary Relationships," p. 18.

155. Tattersall, *The Fossil Trail*, pp. 114–15. See also Stringer, "Evolution of Early Humans," p. 242; Stringer and McKie, *African Exodus*, pp. 32–33.

156. Reader, *Missing Links*, p. 192. See also ibid., pp. 203, 223–26; Stringer, "Evolution of Early Humans," p. 242; Gamble, *Timewalkers*, pp. 54–62.

157. Gamble, *Timewalkers*, p. 54.

158. See, for example, Wood, "Evolution and Australopithecines," p. 231; Groves, "Human Origins," pp. 50–51.

159. Robert Broom, "The Pleistocene Anthropoid Apes of South Africa," in *Naming Our Ancestors: An Anthology of Hominid Taxonomy*, edited by W. Eric Meikle and Sue T. Parker (Prospect Heights, Ill.: Waveland Press, 1994), p. 68. See also p. 65.

160. Gribbin and Cherfas, *The Monkey Puzzle*, p. 28; Goodman, "Reconstructing Human Evolution from Proteins," p. 308.

161. Stringer and McKie, *African Exodus*, p. 20. See also Gribbin and Cherfas, *The Monkey Puzzle*, p. 129.

162. Kelley, "Evolution of Apes," p. 223; Ritvo, "Border Trouble," p. 490.

163. Tattersall, *The Fossil Trail*, pp. 123, 126; Johanson and Edgar, *From Lucy to Language*, p. 40.

164. Diamond, *The Third Chimpanzee*, p. 23.

165. Darwin, *The Descent of Man*, p. 155.

166. Diamond, *The Third Chimpanzee*, pp. 23–25. See also Johanson and Edgar, *From Lucy to Language*, p. 32.

167. George G. Simpson, *Principles of Animal Taxonomy* (New York: Columbia University Press, 1961), pp. 137–39; Tattersall, *The Fossil Trail*, p. 96; Eviatar Zerubavel, "Lumping and Splitting: Notes on Social Classification," *Sociological Forum* 11 (1996): 421–33; Johanson and Edgar, *From Lucy to Language*, p. 52.

168. Bowler, *Theories of Human Evolution*, p. 146. See also Ian Tattersall, "Species Recognition in Human Paleontology," in *Naming Our Ancestors: An Anthology of Hominid Taxonomy*, edited by W. Eric Meikle and Sue T. Parker (Prospect Heights, Ill.: Waveland Press, 1994), pp. 244–45.

169. Ernst Mayr, "Taxonomic Categories in Fossil Hominids," in *Naming Our Ancestors: An Anthology of Hominid Taxonomy*, edited by W. Eric Meikle and Sue T. Parker (Prospect Heights, Ill.: Waveland Press, 1994 [1950]), pp. 152–55, 166; Mayr, "The Taxonomic Evaluation of Fossil Hominids," pp. 377, 382; Louis S. B. Leakey, P. V. Tobias, and J. R. Napier, "A New Species of the Genus *Homo* from Olduvai Gorge," in *Naming Our Ancestors: An Anthology of Hominid Taxonomy*, edited by W. Eric Meikle and Sue T. Parker (Prospect Heights, Ill.: Waveland Press, 1994 [1964]), p. 94.

170. Hammond, "The Expulsion of the Neanderthals from Human Ancestry," p. 3; Tattersall, "Species Recognition in Human Paleontology," pp. 246–50; Gamble, *Timewalkers*, p. 150; Tattersall, *The Fossil Trail*, pp. 219, 231.

171. Tattersall, *The Fossil Trail*, pp. 192–93, 230; Tattersall and Schwartz, *Extinct Humans*, p. 101. See also Tattersall, "Species Recognition in Human Paleontology," p. 247.

172. E. Zerubavel, *The Fine Line*, pp. 78–80. See also Johanna E. Foster, "Menstrual Time: The Sociocognitive Mapping of the Menstrual Cycle," *Sociological Forum* 11 (1996): 523–47; Nicole Isaacson, "The Fetus-Infant: Changing Classifications of *in-utero* Development in Medical Texts," *Sociological Forum* 11 (1996): 457–80.

173. Stocking, "The Persistence of Polygenist Thought in Post-Darwinian Anthropology," p. 61.

174. Johanson and Edgar, *From Lucy to Language*, pp. 18, 32; Friday, "Human Evolution," p. 319; Gould, *Wonderful Life*, p. 29; Gribbin and Cherfas, *The Monkey Puzzle*, p. 129.

175. E. Zerubavel, *The Fine Line*, pp. 64, 79; E. Zerubavel, "Lumping and Splitting," pp. 427–28.

176. Groves, "Human Origins," p. 50; Cavalli-Sforza and Cavalli-Sforza, *The Great Human Diasporas*, pp. 50–55; Gamble, *Timewalkers*, p. 150.

177. Desmond Morris, *The Naked Ape* (New York: McGraw-Hill, 1967); Diamond, *The Third Chimpanzee*.

178. Raymond A. Dart, "Australopithecus Africanus: The Man-Ape of South Africa," in *Naming Our Ancestors: An Anthology of Hominid Taxonomy*, edited by W. Eric Meikle and Sue T. Parker (Prospect Heights, Ill.: Waveland Press, 1994), p. 62.

179. See also Bowler, *Theories of Human Evolution*, p. 29.

180. Dart, "Australopithecus Africanus," pp. 55, 62. See also Matt Cartmill, " 'Four Legs Good, Two Legs Bad': Man's Place (if Any) in Nature," *Natural History* 92 (November 1983): 69.

181. Darwin, *The Descent of Man*, p. 159.

182. E. Zerubavel, *The Fine Line*, pp. 16–17, 76–79; Kristen Purcell, "Leveling the Playing Field: Constructing Parity in the Modern World" (Ph.D. diss., Rutgers University, 2001).

Chapter Four

1. See also Maurice Halbwachs, *The Collective Memory* (New York: Harper Colophon, 1980), pp. 80–82; Eviatar Zerubavel, *The Fine Line: Making Distinctions in Everyday Life* (New York: Free Press, 1991), pp. 9–10, 18–20, 22–23.

2. See also Dan P. McAdams, *The Stories We Live By: Personal Myths and the Making of the Self* (New York: William Morrow, 1993), pp. 256–57.

3. Lisa H. Malkki, *Purity and Exile: Violence, Memory, and National Cosmology among Hutu Refugees in Tanzania* (Chicago: University of Chicago Press, 1995), p. 58.

4. Noel Annan, "Between the Acts," *New York Review of Books*, 24 April 1997, p. 55.

5. See also Yael Zerubavel, *Recovered Roots: Collective Memory and the Making of Israeli National Tradition* (Chicago: University of Chicago Press, 1995), pp. 221–28.

6. John A. Robinson, "First Experience Memories: Contexts and Functions in Personal Histories," in *Theoretical Perspectives on Autobiographical Memory*, edited by Martin A. Conway et al. (Dordrecht: Kluwer Academic Publishers, 1992), p. 225.

7. See also Anselm L. Strauss, *Mirrors and Masks: The Search for Identity* (London: Martin Robertson, 1977), p. 93; Wendy Traas, "Turning Points and Defining Moments: An Exploration of the Narrative Styles That Structure the Personal and Group Identities of Born-Again Christians and Gays and Lesbians" (unpublished manuscript, Rutgers University, Department of Sociology, 2000).

8. Eviatar Zerubavel, *Patterns of Time in Hospital Life: A Sociological Perspective* (Chicago: University of Chicago Press, 1979), p. 92.

9. E. Zerubavel, *The Fine Line*, p. 10; Eviatar Zerubavel, "Language and Memory: 'Pre-Columbian' America and the Social Logic of Periodization," *Social Research* 65 (1998): 318.

10. Emile Durkheim, *The Elementary Forms of Religious Life* (New York: Free Press, 1995 [1912]), p. 313 (page citations are to the reprint edition); Eviatar Zerubavel, *Hidden Rhythms: Schedules and Calendars in Social Life* (Chicago: University of Chicago Press, 1981), pp. 103–5.

11. Israel Bartal, " 'Old Yishuv' and 'New Yishuv': The Image and the Reality," in *Exile in the Homeland: The Settlement of the Land of Israel before Zionism* (in Hebrew) (Jerusalem: Hassifriya Hatziyonit, 1994 [1977]), p. 75 (page citations are to the reprint edition); Yehoshua Kaniel, *Continuity and Change: Old Yishuv and New Yishuv during the First and Second Aliyah* (in Hebrew) (Jerusalem: Yad Itzhak Ben-Zvi Publications, 1981), pp. 21–22; Hana Herzog, "The Concepts 'Old Yishuv' and 'New Yishuv' from a Sociological Perspective" (in Hebrew), *Katedra* 32 (July 1984): 99–108.

12. Jan Assmann, *Moses the Egyptian: The Memory of Egypt in Western Monotheism* (Cambridge, Mass.: Harvard University Press, 1997), pp. 3–8, 208–9.

13. E. Zerubavel, "Language and Memory," p. 323.

14. E. Zerubavel, *The Fine Line*, pp. 16–17, 21–32; Eviatar Zerubavel, "Lumping and Splitting: Notes on Social Classification," *Sociological Forum* 11 (1996): 422–26.

15. See also E. Zerubavel, "Lumping and Splitting," pp. 423–26.

16. Y. Zerubavel, *Recovered Roots*, pp. 17–22.

17. E. Zerubavel, "Language and Memory," pp. 320–21.

18. Y. Zerubavel, *Recovered Roots*, pp. 17–22.

19. E. Zerubavel, *The Fine Line*, pp. 22–23.

20. Harold Rosenberg, *Saul Steinberg* (New York: Alfred A. Knopf, 1978), pp. 62–63, 132. See also Heinz Werner, *Comparative Psychology of Mental Development*, rev. ed. (New York: International Universities Press, 1957), p. 187.

21. E. Zerubavel, "Language and Memory," pp. 321–24.

22. Halbwachs, *The Collective Memory*, p. 101; E. Zerubavel, *Patterns of Time in Hospital Life*, pp. 32–33; Eviatar Zerubavel, *The Seven-Day Circle: The History and Meaning of the Week* (New York: Free Press, 1985), pp. 121, 128.

23. See also E. Zerubavel, *The Fine Line*, pp. 24, 28–32.

24. Ibid., p. 31; Wayne Brekhus, "Social Marking and the Mental Coloring of Identity: Sexual Identity Construction and Maintenance in the United States," *Sociological Forum* 11 (1996): 512. See also Jamie Mullaney, "Like A Virgin: Temptation, Resistance, and the Construction of Identities Based on 'Not Doings,' " *Qualitative Sociology* 24 (2001): 3–24.

25. E. Zerubavel, "Language and Memory," p. 321.

26. *New York Times*, 4 August 2000, sec. A, p. 24.

27. Clifton Daniel, *The Twentieth Century Day by Day* (London: Dorling Kindersley, 2000), p. 782.

28. David C. Gordon, *Self-Determination and History in the Third World* (Princeton, N.J.: Princeton University Press, 1971), pp. 90–96.

29. Arnold Van Gennep, *The Rites of Passage* (Chicago: University of Chicago Press, 1960), p. 11; E. Zerubavel, *The Fine Line*, pp. 23–24. See also Victor Turner, "Betwixt and Between: The Liminal Period in Rites de Passage," in *The Forest of Symbols: Aspects of Ndembu Ritual* (Ithaca, N.Y.: Cornell University Press, 1970), pp. 93–111.

30. Richard I. Jobs, "The Promise of Youth: Age Categories as the Mental Framework of Rejuvenation in Postwar France" (paper presented at the Tenth Annual Interdisciplinary Conference for Graduate Scholarship, the Center for the Critical Analysis of Contemporary Culture, Rutgers University, New Brunswick, N.J., March 2000).

31. Y. Zerubavel, *Recovered Roots*, pp. 20–31; Oz Almog, *The Sabra: The Creation of the New Jew* (Berkeley and Los Angeles: University of California Press, 2000); Yael Zerubavel, "The Mythological Sabra and the Jewish Past: Trauma, Memory, and Contested Identities," *Israel Studies* 7 (2002).

32. See also Eviatar Zerubavel, "The French Republican Calendar: A Case Study in the Sociology of Time," *American Sociological Review* 42 (1977): 868–77; Eviatar Zerubavel, "Easter and Passover: On Calendars and Group Identity," *American Sociological Review* 47 (1982): 284–89.

33. Karen A. Cerulo, *Identity Designs: The Sights and Sounds of a Nation* (New Brunswick, N.J.: Rutgers University Press, 1995), pp. 154–65.

34. Maoz Azaryahu, "The Purge of Bismarck and Saladin: The Renaming of Streets in East Berlin and Haifa," *Poetics Today* 13 (1992): 351–66.

35. See also E. Zerubavel, *Patterns of Time in Hospital Life*, p. 92.

36. Samuel L. Clemens [Mark Twain, pseud.], *Life on the Mississippi* (New York: Magnum Easy Eye Books, 1968), p. 389. Emphasis added.

37. Pitirim A. Sorokin and Robert K. Merton, "Social Time: A Methodological and Functional Analysis," *American Journal of Sociology* 42 (1937): 623; Pitirim A. Sorokin, *Sociocultural Causality, Space, Time: A Study of Referential Principles of Sociology and Social Science* (Durham, N.C.: Duke University Press, 1943), p. 174; E. Zerubavel, *Hidden Rhythms*, pp. 86–87; Eviatar Zerubavel, "In the Beginning: Notes on the Social Construction of Historical Discontinuity," *Sociological Inquiry* 63 (1993): 457–58.

38. Kenneth L. Woodward, "2000 Years of Jesus," *Newsweek*, 29 March 1999, p. 54.

39. See also Bernard Lewis, *History: Remembered, Recovered, Invented* (Princeton, N.J.: Princeton University Press, 1975), p. 32.

40. See also Mullaney, "Like A Virgin," p. 10.

41. Claudia Koonz, "Between Memory and Oblivion: Concentration Camps in German Memory," in *Commemorations: The Politics of National Identity*, edited by John R. Gillis (Princeton, N.J.: Princeton University Press, 1994), p. 262.

42. Ruth W. Gregory, *Anniversaries and Holidays*, 4th ed. (Chicago: American Library Association, 1983), p. 53.

43. George G. Andrews, "Making the Revolutionary Calendar," *American Historical Review* 36 (1931): 517–23; E. Zerubavel, *Hidden Rhythms*, pp. 86–87.

44. Herbert W. Schneider and Shepard B. Clough, *Making Fascists* (Chicago: University of Chicago Press, 1929), p. 193; Mabel Berezin, *Making the Fascist Self: The Political Culture of Interwar Italy* (Ithaca, N.Y.: Cornell University Press, 1997), p. 67.

45. *Facts on File*, vol. 39 (1979), p. 247.

46. See Y. Zerubavel, *Recovered Roots*, pp. 15–22.

47. Yael Zerubavel, *Desert Images: Visions of the Counter-Place in Israeli Culture* (Chicago: University of Chicago Press, forthcoming).

48. On mental "nearsightedness" and "farsightedness," see Ruth Simpson, "Microscopic Worlds, Miasmatic Theories, and Myopic Vision: Changing Conceptions of Air and Social Space" (paper presented at the Annual Meeting of the American Sociological Association, Chicago, 1999).

49. See Gwyn Jones, *The Norse Atlantic Saga*, 2d ed. (Oxford: Oxford University Press, 1986), pp. 144, 156.

50. See also Martin W. Lewis and Kären E. Wigen, *The Myth of Continents: A Critique of Metageography* (Berkeley and Los Angeles: University of California Press, 1997), pp. 106–11.

51. E. Zerubavel, "Language and Memory," pp. 320–22.

52. Henry Glassie, *Passing the Time in Ballymenone: Culture and History of an Ulster Community* (Philadelphia: University of Pennsylvania Press, 1982), p. 626.

53. Noam Chomsky, *Year 501: The Conquest Continues* (Boston: South End Press, 1993).

54. Eviatar Zerubavel, *Terra Cognita: The Mental Discovery of America* (New Brunswick, N.J.: Rutgers University Press, 1992), pp. 13–23, 26–28.

55. Amnon Rubinstein, *To Be a Free People* (in Hebrew) (Tel-Aviv: Schocken, 1977), p. 104.

56. See, for example, Y. Zerubavel, *Recovered Roots*, p. xv. But see Y. Zerubavel, "The Mythological Sabra and the Jewish Past."

57. See also James W. Loewen, *Lies My Teacher Told Me: Everything Your American History Textbook Got Wrong* (New York: Touchstone, 1996), p. 77.

58. Berl Katznelson, quoted in Doron Rosenblum, "Because Somebody Needs to Be an Israeli in Israel" (in Hebrew), *Ha'aretz* (Independence Day supplement), 29 April 1998, p. 52.

59. Eviatar Zerubavel, *Social Mindscapes: An Invitation to Cognitive Sociology* (Cambridge, Mass.: Harvard University Press, 1997), p. 85.

60. Ibid., pp. 84–85; E. Zerubavel, "Language and Memory," pp. 318–26. On the social construction of irrelevance, see also Eviatar Zerubavel, "The Elephant in the Room: Notes on the Social Organization of Denial," in *Culture in Mind: Toward a Sociology of Culture and Cognition*, edited by Karen A. Cerulo (New York: Routledge, 2002), pp. 21–27.

61. E. Zerubavel, *Social Mindscapes*, pp. 84–85.

62. Roger Cohen, "Anniversary Sets Germans to Quarreling on Holocaust," *New York Times*, 10 November 1998, International section, p. A16.

63. "Montenegro Asks Forgiveness from Croatia," *New York Times*, 25 June 2000, International section, p. 9.

64. Seth Mydans, "Under Prodding, 2 Apologize for Cambodian Anguish," *New York Times*, 30 December, 1998, International section.

65. Seth Mydans, "Cambodian Leader Resists Punishing Top Khmer Rouge," *New York*

Times, 29 December 1998, sec. A, p. 1. Emphasis added.

66. See, for example, Edward E. Cummings, *Complete Poems* (New York: Harcourt Brace Jovanovich, 1972); Erving Goffman, *Frame Analysis: An Essay on the Organization of Experience* (New York: Harper Colophon, 1974), p. 391 n.

67. Netiva Ben-Yehuda, *1948: Between the Eras* (in Hebrew) (Jerusalem: Keter, 1981).

68. Jacob Barnai, *Historiography and Nationalism: Trends in the Research of Palestine and Its Jewish Yishuv, 634–1881* (in Hebrew) (Jerusalem: Magnes Press, 1995), pp. 180, 188.

69. Elliott West, "A Longer, Grimmer, but More Interesting Story," in *Trails toward A New Western History*, edited by Patricia Nelson Limerick et al. (Lawrence: University Press of Kansas, 1991), p. 107.

70. E. Zerubavel, "Language and Memory," pp. 324–25.

71. George Kubler, *The Shape of Time* (New Haven, Conn.: Yale University Press, 1962), p. 56.

72. See also E. Zerubavel, *The Fine Line*, p. 72.

73. See also ibid., pp. 61, 116.

74. See also Johanna E. Foster, "Menstrual Time: The Sociocognitive Mapping of 'The Menstrual Cycle,'" *Sociological Forum* 11 (1996): 533–36, 541–42.

75. Christopher Stringer and Robin McKie, *African Exodus: The Origins of Modern Humanity* (New York: Henry Holt, 1997), pp. 261–62. See also, in this regard, Nadia Abu El-Haj, *Facts on the Ground: Archaeological Practice and Territorial Self-Fashioning in Israeli Society* (Chicago: University of Chicago Press, 2001).

76. See also Dietrich Gerhard, "Periodization in History," in *Dictionary of the History of Ideas: Studies of Selected Pivotal Ideas*, edited by Philip P. Wiener (New York: Charles Scribner's Sons, 1973), 3:476.

77. Dietrich Gerhard, "Periodization in European History," *American Historical Review* 61 (1956): 901; Wallace K. Ferguson, *The Renaissance* (New York: Henry Holt, 1940), p. 1. See also Gerhard, "Periodization in History," p. 479.

78. Gerhard, "Periodization in History," p. 477.

79. Jason S. Smith, "The Strange History of the Decade: Modernity, Nostalgia, and the Perils of Periodization," *Journal of Social History* 32 (1998): 269–75. See also Fred Davis, "Decade Labeling: The Play of Collective Memory and Narrative Plot," *Symbolic Interaction* 7, no. 1 (1984): 15–24.

80. E. Zerubavel, *The Fine Line*, p. 76.

81. See also Nicole Isaacson, "The Fetus-Infant: Changing Classifications of *in-utero* Development in Medical Texts," *Sociological Forum* 11 (1996): 472–76; Foster, "Menstrual Time," pp. 531–35.

82. See also Isaacson, "The Fetus-Infant," pp. 467–70.

83. Clifford Geertz, "The Impact of the Concept of Culture on the Concept of Man," in *The Interpretation of Cultures* (New York: Basic Books, 1973), p. 47.

84. Richard Dawkins, *River out of Eden: A Darwinian View of Life* (New York: Basic Books, 1995), pp. 9–10.

85. See, for example, Misia Landau, "Human Evolution as Narrative," *American Scientist* 72 (1984): 262–68; Ian Tattersall, *The Fossil Trail: How We Know What We Think We Know about Human Evolution* (New York: Oxford University Press, 1995), pp. 114–15; Stringer and McKie, *African Exodus*, pp. 32–33; Ian Tattersall and Jeffrey H. Schwartz, *Extinct Humans* (Boulder, Colo.: Westview, 2000), pp. 236–38; John N.

Wilford, "When Humans Became Human," *New York Times,* 26 February 2002, sec. F, pp. 1, 5; Natalie Angier, "Cooking and How It Slew the Beast Within," *New York Times,* 28 May 2002, sec. F, pp. 1–6.

86. Barry S. Strauss, "The Problem of Periodization: The Case of the Peloponnesian War," in *Inventing Ancient Culture: Historicism, Periodization, and the Ancient World,* edited by Mark Golden and Peter Toohey (London: Routledge, 1997), pp. 165–75.

87. E. Zerubavel, *Patterns of Time in Hospital Life,* pp. 99–100; Foster, "Menstrual Time," pp. 538–39.

88. See also J. H. Hexter, *On Historians: Reappraisals of Some of the Makers of Modern History* (Cambridge, Mass.: Harvard University Press, 1979), pp. 242–43.

89. See Zeev Schiff and Eitan Haber, eds., *Israel, Army, and Defense: A Dictionary* (in Hebrew) (Jerusalem: Zmora, Bitan, Modan, 1976), pp. 15, 182.

90. *Ha'aretz* weekly supplement, 13 April 2001.

91. Michael Maclear, *The Ten Thousand Day War: Vietnam, 1945–1975* (Toronto: Methuen, 1981).

92. See also Marilyn B. Young, *The Vietnam Wars 1945–1990* (New York: HarperCollins, 1991); Strauss, "The Problem of Periodization," p. 165.

93. Ian Buruma, *The Wages of Guilt: Memories of War in Germany and Japan* (New York: Farrar Straus Giroux, 1994), p. 48.

94. Alistair Horne, *To Lose a Battle: France 1940* (Boston: Little, Brown & Co., 1969), p. 584.

95. James Bennet, "Year of Intifada Sees Hardening on Each Side," *New York Times,* 28 September 2001, sec. A, p. 3.

96. See, for example, Bill Powell, "The Innocents of WW II?" *Newsweek,* 12 December 1994, p. 53; Roy F. Baumeister and Stephen Hastings, "Distortions of Collective Memory: How Groups Flatter and Deceive Themselves," in *Collective Memory of Political Events: Social Psychological Perspectives,* edited by James W. Pennebaker et al. (Mahwah, N.J.: Lawrence Erlbaum, 1997), pp. 284–85.

97. Don Van Natta and James Risen, "Investigators Say Evidence Points to Bin Laden Aides as Planning Attack," *New York Times,* 8 October 2001, sec. B, p. 7.

98. See Noel Malcolm, *Kosovo: A Short History,* updated ed. (New York: HarperPerennial, 1999), pp. 139–62, 251–52, 327.

Chapter Five

1. David B. Pillemer, *Momentous Events, Vivid Memories* (Cambridge, Mass.: Harvard University Press, 1998), pp. 88–91.

2. Lyn Spillman, *Nation and Commemoration: Creating National Identities in the United States and Australia* (Cambridge: Cambridge University Press, 1997).

3. See also Michael A. Katovich and Carl J. Couch, "The Nature of Social Pasts and Their Use as Foundations for Situated Action," *Symbolic Interaction* 15, no. 1 (1992): 25–47.

4. Alex Shoumatoff, *The Mountain of Names: A History of the Human Family* (New York: Simon and Schuster, 1985), pp. 62–63, 69, 78.

5. Alex Haley, *Roots* (Garden City, N.Y.: Doubleday, 1976), p. 1.

6. See also Meyer Fortes, "The Significance of Descent in Tale Social Structure," in *Time and Social Structure and Other Essays* (London: Athlone Press, 1970), p. 41.

7. Bernard Lewis, *History: Remembered, Recovered, Invented* (Princeton, N.J.: Princeton University Press, 1975), p. 101. See also p. 41.

8. See, for example, Mary C. Waters, *Ethnic Options: Choosing Identities in America* (Berkeley and Los Angeles: University of California Press, 1990); Kwame A. Appiah, *In My Father's House: Africa in the Philosophy of Culture* (New York: Oxford University Press, 1992), p. viii; Johanna E. Foster, "Feminist Theory and the Politics of Ambiguity: A Comparative Analysis of the Multiracial Movement, the Intersex Movement and the Disability Rights Movement as Contemporary Struggles over Social Classification in the United States" (Ph.D. diss., Rutgers University, 2000).

9. Appiah, *In My Father's House*, p. 32. See also pp. 28–31.

10. B. Lewis, *History*, pp. 31–41.

11. David C. Gordon, *Self-Determination and History in the Third World* (Princeton, N.J.: Princeton University Press, 1971), p. 98; Michael Herzfeld, *Ours Once More: Folklore, Ideology, and the Making of Modern Greece* (New York: Pella, 1986), p. 40; Israel Gershoni and James P. Jankowski, *Egypt, Islam, and the Arabs: The Search for Egyptian Nationhood, 1900–1930* (New York: Oxford University Press, 1986), pp. 143–63.

12. Gordon, *Self-Determination and History in the Third World*, pp. 90–91, 102–3.

13. B. Lewis, *History*, p. 34.

14. Lisa H. Malkki, *Purity and Exile: Violence, Memory, and National Cosmology among Hutu Refugees in Tanzania* (Chicago: University of Chicago Press, 1995), pp. 59–61.

15. David Lowenthal, *Possessed by the Past: The Heritage Crusade and the Spoils of History* (New York: Free Press, 1996), pp. 173–91.

16. *New York Times*, 15 June 2001, sec. A, p. 8. Emphasis added.

17. Celestine Bohlen, "In Transylvania, the Battle for the Past Continues," *New York Times*, 18 March 1990, International section, p. 16. See also, in this regard, Nadia Abu El-Haj, *Facts on the Ground: Archaeological Practice and Territorial Self-Fashioning in Israeli Society* (Chicago: University of Chicago Press, 2001).

18. Israel Bartal, "Invented First Aliya: How Counter-History Works in Religious Zionism" (paper presented at the Annual Meeting of the Association for Jewish Studies, Chicago, December 1999).

19. Michael Z. Wise, "Idea of a Unified Cultural Heritage Divides Europe," *New York Times*, 29 January 2000, sec. B, pp. 9–11.

20. Michael T. Kaufman, "Two Distinct Peoples with Two Divergent Memories Battle over One Land," *New York Times*, 4 April 1999, International section, p.10. See also Noel Malcolm, *Kosovo: A Short History*, updated ed. (New York: HarperPerennial, 1999), pp. 22–23, 28–30.

21. Yael Zerubavel, *Recovered Roots: Collective Memory and the Making of Israeli National Tradition* (Chicago: University of Chicago Press, 1995), pp. 15–36.

22. Jeffrey Goldberg, "Israel's Y2K Problem," *New York Times Magazine*, 3 October 1999, p. 77.

23. Rashid I. Khalidi, "What 'Final Status'?" *New York Times*, 3 October 1996, sec. A, p. 23.

24. Gary A. Rendsburg, "Archeological Fairness," *New York Times*, 11 October 1996, sec. A, p. 38. Emphasis added.

25. Lyn Spillman, "When Do Collective Memories Last? Founding Moments in the United States and Australia," *Social Science History* 22 (1998): 463.

26. David Remnick, "The Outsider," *The New Yorker*, 25 May 1998, pp. 86–88. Emphasis added.

27. Karl Mannheim, *Ideology and Utopia: An Introduction to the Sociology of Knowledge* (New York: Harvest Books, 1936), pp. 12, 33.

Bibliography

Abbott, Andrew. "On the Concept of the Turning Point." *Comparative Social Research* 16 (1997): 85–105.

Abbott, Edwin A. *Flatland: A Romance of Many Dimensions*. New York: Dover, 1992. Originally published in 1884.

Abu El-Haj, Nadia. *Facts on the Ground: Archaeological Practice and Territorial Self-Fashioning in Israeli Society*. Chicago: University of Chicago Press, 2001.

Allen, Frederick. "They're Still There: The Oldest Business in America." *American Heritage of Invention and Technology* 15, no. 3 (2000): 6.

Almog, Oz. *The Sabra: The Creation of the New Jew*. Berkeley and Los Angeles: University of California Press, 2000. Originally published in 1997.

Andrews, George G. "Making the Revolutionary Calendar." *American Historical Review* 36 (1931): 515–32.

Angier, Natalie. "Do Races Differ? Not Really, Genes Show." *New York Times*, 22 August 2000, sec. F, pp. 1–6.

————. "Cooking and How It Slew the Beast Within." *New York Times*, 28 May 2002, sec. F, pp. 1–6.

Annan, Noel. "Between the Acts." *New York Review of Books*, 24 April 1997, pp. 55–59.

Appiah, Kwame A. *In My Father's House: Africa in the Philosophy of Culture*. New York: Oxford University Press, 1992.

Apple, R. W. Jr. "A Military Quagmire Remembered: Afghanistan as Vietnam." *New York Times*, 31 October 2001, sec. B, pp. 1, 3.

Assmann, Jan. *Moses the Egyptian: The Memory of Egypt in Western Monotheism*. Cambridge, Mass.: Harvard University Press, 1997.

Ayoub, Millicent R. "The Family Reunion." *Ethnology* 5 (1966): 415–33.

Azaryahu, Maoz. "The Purge of Bismarck and Saladin: The Renaming of Streets in East Berlin and Haifa." *Poetics Today* 13 (1992): 351–66.

Babylonian Talmud, The. London: Soncino, 1938.

Baker, Hugh. *Chinese Family and Kinship*. New York: Columbia University Press, 1979.

Balázs, Béla. *Theory of the Film: Character and Growth of a New Art.* New York: Dover, 1970. Originally published in 1945.

Barnai, Jacob. *Historiography and Nationalism: Trends in the Research of Palestine and Its Jewish Yishuv, 634–1881* (in Hebrew). Jerusalem: Magnes Press, 1995.

Bartal, Israel. " 'Old Yishuv' and 'New Yishuv': The Image and the Reality." In *Exile in the Homeland: The Settlement of the Land of Israel before Zionism* (in Hebrew), pp. 74–89. Jerusalem: Hassifriya Hatziyonit, 1994. Originally published in 1977.

———. "Invented First Aliya: How Counter-History Works in Religious Zionism." Paper presented at the Annual Meeting of the Association for Jewish Studies, Chicago, December 1999.

Bartlett, Frederic C. *Remembering: A Study in Experimental and Social Psychology.* Cambridge: Cambridge University Press, 1932.

Baumeister, Roy F., and Stephen Hastings. "Distortions of Collective Memory: How Groups Flatter and Deceive Themselves." In *Collective Memory of Political Events: Social Psychological Perspectives,* edited by James W. Pennebaker et al., pp. 277–93. Mahwah, N.J.: Lawrence Erlbaum, 1997.

Belluck, Pam. "Pilgrims Wear Different Hats in Recast Thanksgiving Tales." *New York Times,* 23 November 1995, sec. A, p. 1; sec. B, p. 7.

Bennet, James. "Hillary Clinton, in Morocco, Says NATO Attack Aims at Stopping Bloodshed." *New York Times,* 31 March 1999, International section, p. A10.

———. "Year of Intifada Sees Hardening on Each Side." *New York Times,* 28 September 2001, sec. A., p. 3.

———. "Sharon Invokes Munich in Warning U.S. on 'Appeasement.' " *New York Times,* 5 October 2001, sec. A, p. 6.

Ben-Simon, Daniel. "A Secure Step in a Sealed City" (in Hebrew). *Ha'aretz,* 28 August 1998, p. 14.

———. "The Settlers' Nightmares" (in Hebrew). *Ha'aretz,* 23 June 2000, p. 16.

Ben-Yehuda, Nachman. *The Masada Myth: Collective Memory and Mythmaking in Israel.* Madison: University of Wisconsin Press, 1995.

Ben-Yehuda, Netiva. *1948: Between the Eras* (in Hebrew). Jerusalem: Keter, 1981.

Berezin, Mabel. *Making the Fascist Self: The Political Culture of Interwar Italy.* Ithaca, N.Y.: Cornell University Press, 1997.

Berger, Peter L. *Invitation to Sociology: A Humanistic Perspective.* Garden City, N.Y.: Doubleday Anchor, 1963.

Berger, Peter L., and Thomas Luckmann. *The Social Construction of Reality: A Treatise in the Sociology of Knowledge.* Garden City, N.Y.: Doubleday, 1966.

Bergson, Henri. *Time and Free Will: An Essay on the Immediate Data of Consciousness.* New York: Harper and Row, 1960. Originally published in 1889.

Bernstein, Michael A. *Foregone Conclusions: Against Apocalyptic History.* Berkeley and Los Angeles: University of California Press, 1994.

Blackburn, Gilmer W. *Education in the Third Reich: Race and History in Nazi Textbooks.* Albany: State University of New York Press, 1985.

Bohlen, Celestine. "In Transylvania, the Battle for the Past Continues." *New York Times,* 18 March 1990, International section, p. 16.

Böröcz, József. "Sticky Features: Narrating a Single Direction." Paper presented at the "Beginnings and Endings" seminar at the Center for the Critical Analysis of Contemporary Culture, Rutgers University, New Brunswick, N.J., September 1999.

Bouquet, Mary. "Family Trees and Their Affinities: The Visual Imperative of the Genealogical Diagram." *Journal of the Royal Anthropological Institute*, n.s., 2 (1996): 43–66.

Bowler, Peter J. *Theories of Human Evolution: A Century of Debate, 1844–1944*. Baltimore: Johns Hopkins University Press, 1986.

———. *The Invention of Progress: The Victorians and the Past*. Oxford: Basil Blackwell, 1989.

———. *Life's Splendid Drama: Evolutionary Biology and the Reconstruction of Life's Ancestry, 1860–1940*. Chicago: University of Chicago Press, 1996.

Bragg, Rick. "Emotional March Gains a Repentant Wallace." *New York Times*, 11 March 1995, sec. A, pp. 1, 9.

Braude, Benjamin. "The Sons of Noah and the Construction of Ethnic and Geographical Identities in the Medieval and Early Modern Periods." *The William and Mary Quarterly*, 3d ser., 54 (1997): 103–42.

Brekhus, Wayne. "Social Marking and the Mental Coloring of Identity: Sexual Identity Construction and Maintenance in the United States." *Sociological Forum* 11 (1996): 497–522.

Bronowski, Jacob. *The Ascent of Man*. Boston: Little, Brown & Co., 1973.

Brooke, James. "Conquistador Statue Stirs Hispanic Pride and Indian Rage." *New York Times*, 9 February 1998, sec. A, p. 10.

Broom, Robert. "The Pleistocene Anthropoid Apes of South Africa." In *Naming Our Ancestors: An Anthology of Hominid Taxonomy*, edited by W. Eric Meikle and Sue T. Parker, pp. 65–70. Prospect Heights, Ill.: Waveland Press, 1994. Originally published in 1938.

Bruni, Frank, and Katharine Q. Seelye. "Campaign Contrasts Grow Starker." *New York Times*, 2 July 2000, sec. A, p. 11.

Brzezinski, Zbigniew. "Can Communism Compete with the Olympics?" *New York Times*, 14 July 2001, sec. A, p. 15.

Burenhult, Göran. "Modern People in Africa and Europe." In *The First Humans: Human Origins and History to 10,000 B.C.*, edited by Göran Burenhult, pp. 77–95. New York: HarperCollins, 1993.

Burns, John F. "New Babylon Is Stalled by a Modern Upheaval." *New York Times*, 11 October 1990, International section, p. A13.

Buruma, Ian. *The Wages of Guilt: Memories of War in Germany and Japan*. New York: Farrar Straus Giroux, 1994.

Cann, Rebecca L., Mark Stoneking, and Allan C. Wilson. "Mitochondrial DNA and Human Evolution." *Nature* 325 (1987): 31–36.

Cartmill, Matt. " 'Four Legs Good, Two Legs Bad': Man's Place (if Any) in Nature." *Natural History* 92 (November 1983): 64–79.

Cavalli-Sforza, Luigi L. *Genes, Peoples, and Languages*. New York: North Point Press, 2000.

Cavalli-Sforza, Luigi L., and Francesco Cavalli-Sforza. *The Great Human Diasporas: The History of Diversity and Evolution*. Reading, Mass.: Addison-Wesley, 1995. Originally published in 1993.

Cavalli-Sforza, Luigi L., Paolo Menozzi, and Alberto Piazza. *The History and Geography of Human Genes.* Abridged pbk ed. Princeton, N.J.: Princeton University Press, 1996.

Cerulo, Karen A. *Identity Designs: The Sights and Sounds of a Nation.* New Brunswick, N.J.: Rutgers University Press, 1995.

Cerulo, Karen A., and Janet M. Ruane. "Death Comes Alive: Technology and the Reconception of Death." *Science as Culture* 6 (1997): 444–66.

Chamberlin, E. R. *Preserving the Past.* London: J. M. Dent, 1979.

Chambers, Robert. *Vestiges of the Natural History of Creation.* Chicago: University of Chicago Press, 1994. Originally published in 1844.

Chancey, Matthew L. "Mrs. Alberta Martin: The Old Man's Darling." <http://lastconfederatewidow.com>, accessed 7 February 2002.

Chase's 1997 Calendar of Events. Chicago: Contemporary Publishing Co., 1996.

Chomsky, Noam. *Year 501: The Conquest Continues.* Boston: South End Press, 1993.

Clanchy, M. T. *From Memory to Written Record: England, 1066–1307.* Cambridge, Mass.: Harvard University Press, 1979.

Clemens, Samuel L. [Mark Twain, pseud.]. *Life on the Mississippi.* New York: Magnum Easy Eye Books, 1968. Originally published in 1883.

Cohen, Roger. "Anniversary Sets Germans to Quarreling on Holocaust." *New York Times,* 10 November 1998, International section, p. A16.

Collins, Randall. *The Sociology of Philosophies: A Global Theory of Intellectual Change.* Cambridge, Mass.: Harvard University Press, 1998.

Comte, Auguste. *Cours de Philosophie Positive.* In *Auguste Comte and Positivism: The Essential Writings,* edited by Gertrud Lenzer, pp. 71–306. New York: Harper Torchbooks, 1975. Originally published in 1830–42.

Connerton, Paul. *How Societies Remember.* Cambridge: Cambridge University Press, 1989.

Coon, Carleton S. *The Origin of Races.* New York: Alfred A. Knopf, 1962.

Cooper, Nancy, and Christopher Dickey. "After the War: Iraq's Designs." *Newsweek,* 8 August 1988, pp. 34–35.

Coser, Lewis A., and Rose L. Coser. "Time Perspective and Social Structure." In *Modern Sociology: An Introduction to the Science of Human Interaction,* edited by Alvin W. Gouldner and Helen P. Gouldner, pp. 638–47. New York: Harcourt, Brace & World, 1963.

Crocker, Lester. "Diderot and Eighteenth Century French Transformism." In *Forerunners of Darwin: 1745–1859,* edited by Bentley Glass et al., pp. 114–43. Baltimore: Johns Hopkins University Press, 1959.

Cummings, Edward E. *Complete Poems.* New York: Harcourt Brace Jovanovich, 1972.

Daniel, Clifton. *Chronicle of America.* Mount Kisco, N.Y.: Chronicle Publications, 1989.

———. *The Twentieth Century Day by Day.* London: Dorling Kindersley, 2000.

Dankner, Amnon, and David Tartakover. *Where We Were and What We Did: An Israeli Lexicon of the Fifties and the Sixties* (in Hebrew). Jerusalem: Keter, 1996.

Dart, Raymond A. "Australopithecus Africanus: The Man-Ape of South Africa." In *Naming Our Ancestors: An Anthology of Hominid Taxonomy,* edited by W. Eric Meikle and Sue T. Parker, pp. 53–70. Prospect Heights, Ill.: Waveland Press, 1994. Originally published in 1925.

Darwin, Charles. *The Origin of Species*. New York: Mentor Books, 1958. Originally published in 1859.

———. *The Descent of Man and Selection in Relation to Sex*. Amherst, N.Y.: Prometheus, 1998. Originally published in 1871.

Davis, Eric. "The Museum and the Politics of Social Control in Modern Iraq." In *Commemorations: The Politics of National Identity*, edited by John R. Gillis, pp. 90–104. Princeton, N.J.: Princeton University Press, 1994.

Davis, Fred. *Yearning for Yesterday: A Sociology of Nostalgia*. New York: Free Press, 1979.

———. "Decade Labeling: The Play of Collective Memory and Narrative Plot." *Symbolic Interaction* 7, no. 1 (1984): 15–24.

Davis, Murray S. *Smut: Erotic Reality/Obscene Ideology*. Chicago: University of Chicago Press, 1983.

Dawkins, Richard. *River out of Eden: A Darwinian View of Life*. New York: Basic Books, 1995.

DePalma, Anthony. "In the War Cry of the Indians, Zapata Rides Again." *New York Times*, 27 January 1994, International section.

Desmond, Adrian. *Archetypes and Ancestors: Palaeontology in Victorian London 1850–1875*. Chicago: University of Chicago Press, 1984. Originally published in 1982.

Diamond, Jared. *The Third Chimpanzee: The Evolution and Future of the Human Animal*. New York: HarperCollins, 1992.

Domínguez, Virginia R. *White by Definition: Social Classification in Creole Louisiana*. New Brunswick, N.J.: Rutgers University Press, 1986.

Douglas, Mary. *How Institutions Think*. Syracuse, N.Y.: Syracuse University Press, 1986.

Dowd, Maureen. "Center Holding." *New York Times*, 20 May 1998, sec. A, p. 23.

Durkheim, Emile. *The Elementary Forms of Religious Life*. New York: Free Press, 1995. Originally published in 1912.

Ebaugh, Helen R. F. *Becoming an Ex: The Process of Role Exit*. Chicago: University of Chicago Press, 1988.

Eldredge, Niles, and Stephen J. Gould. "Punctuated Equilibria: An Alternative to Phyletic Gradualism." In *Models in Paleobiology*, edited by Thomas J. Schopf, pp. 82–115. San Francisco: Freeman, Cooper, & Co., 1972.

Eliade, Mircea. *The Sacred and the Profane: The Nature of Religion*. New York: Harcourt, Brace & World, 1959. Originally published in 1957.

Erikson, Kai T. *Everything in Its Path: Destruction of Community in the Buffalo Creek Flood*. New York: Simon and Schuster, 1976.

Europa World Year Book 1997. London: Europa Publications, 1997.

Evans-Pritchard, Edward E. *The Nuer: A Description of the Modes of Livelihood and Political Institutions of a Nilotic People*. London: Oxford University Press, 1940.

Fabian, Johannes. *Time and the Other: How Anthropology Makes Its Object*. New York: Columbia University Press, 1983.

Fain, Haskell. *Between Philosophy and History: The Resurrection of Speculative Philosophy of History within the Analytic Tradition*. Princeton, N.J.: Princeton University Press, 1970.

Faison, Seth. "Not Equal to Confucius, but Friends to His Memory." *New York Times*, 10 October 1997, International section.

Ferguson, Wallace K. *The Renaissance*. New York: Henry Holt, 1940.

Firth, Raymond. "A Note on Descent Groups in Polynesia." In *Kinship and Social Organization*, edited by Paul Bohannan and John Middleton, pp. 213–23. Garden City, N.Y.: American Museum of Natural History, 1968. Originally published in 1957.

Fischer, David H. *The Great Wave: Price Revolutions and the Rhythm of History*. New York: Oxford University Press, 1996.

Fivush, Robyn, Catherine Haden, and Elaine Reese. "Remembering, Recounting, and Reminiscing: The Development of Autobiographical Memory in Social Context." In *Remembering Our Past: Studies in Autobiographical Memory*, edited by David C. Rubin, pp. 341–58. Cambridge: Cambridge University Press, 1996.

Flaherty, Michael G. *A Watched Pot: How We Experience Time*. New York: New York University Press, 1999.

Forrest, Thomas R. "Disaster Anniversary: A Social Reconstruction of Time." *Sociological Inquiry* 63 (1993): 444–56.

Fortes, Meyer. "The Significance of Descent in Tale Social Structure." In *Time and Social Structure and Other Essays*, pp. 33–66. London: Athlone Press, 1970. Originally published in 1943–44.

———. "Descent, Filiation, and Affinity." In *Time and Social Structure and Other Essays*, pp. 96–121. London: Athlone Press, 1970. Originally published in 1959.

Foster, Johanna E. "Menstrual Time: The Sociocognitive Mapping of 'The Menstrual Cycle.'" *Sociological Forum* 11 (1996): 523–47.

———. "Feminist Theory and the Politics of Ambiguity: A Comparative Analysis of the Multiracial Movement, the Intersex Movement and the Disability Rights Movement as Contemporary Struggles over Social Classification in the United States." Ph.D. diss., Rutgers University, 2000.

Freeman, J. D. "On the Concept of the Kindred." In *Kinship and Social Organization*, edited by Paul Bohannan and John Middleton, pp. 255–72. Garden City, N.Y.: American Museum of Natural History, 1968. Originally published in 1961.

Freud, Sigmund. *Civilization and Its Discontents*. New York: W. W. Norton, 1962. Originally published in 1930.

Friday, Adrian E. "Human Evolution: The Evidence from DNA Sequencing." In *The Cambridge Encyclopedia of Human Evolution*, edited by Steve Jones et al., pp. 316–21. Cambridge: Cambridge University Press, 1992.

Friedberg, Avraham S. *Zikhronot le-Veit David* (in Hebrew). Ramat Gan, Israel: Masada, 1958. Originally published in 1893–1904.

Frisch, Michael. "American History and the Structures of Collective Memory: A Modest Exercise in Empirical Iconography." *Journal of American History* 75 (1989): 1130–55.

Gamble, Clive. *Timewalkers: The Prehistory of Global Colonization*. Cambridge, Mass.: Harvard University Press, 1994.

Gangi, Giuseppe. *Rome Then and Now*. Rome: G & G Editrice, n.d.

Garfinkel, Harold. "Passing and the Managed Achievement of Sex Status in an 'Intersexed' Person." In *Studies in Ethnomethodology*, pp. 116–85. Englewood Cliffs, N.J.: Prentice-Hall, 1967.

Geertz, Clifford. "The Impact of the Concept of Culture on the Concept of Man." In *The*

Interpretation of Cultures, pp. 33–54. New York: Basic Books, 1973. Originally published in 1966.

Gerhard, Dietrich. "Periodization in European History." *American Historical Review* 61 (1956): 900–913.

————. "Periodization in History." In *Dictionary of the History of Ideas: Studies of Selected Pivotal Ideas*, vol. 3, edited by Philip P. Wiener, pp. 476–81. New York: Charles Scribner's Sons, 1973.

Gershoni, Israel, and James P. Jankowski. *Egypt, Islam, and the Arabs: The Search for Egyptian Nationhood, 1900–1930*. New York: Oxford University Press, 1986.

Gillis, John R. *A World of Their Own Making: Myth, Ritual, and the Quest for Family Values.* New York: Basic Books, 1996.

Glassie, Henry. *Passing the Time in Ballymenone: Culture and History of an Ulster Community.* Philadelphia: University of Pennsylvania Press, 1982.

Gobineau, Arthur de. *The Inequality of Human Races*. New York: Howard Fertig, 1967. Originally published in 1854.

Goffman, Erving. *Stigma: Notes on the Management of Spoiled Identity*. Englewood Cliffs, N.J.: Prentice-Hall, 1963.

————. *Frame Analysis: An Essay on the Organization of Experience*. New York: Harper Colophon, 1974.

Goldberg, Carey. "DNA Offers Link to Black History." *New York Times*, 28 August 2000, sec. A, p. 10.

Goldberg, Jeffrey. "Israel's Y2K Problem." *New York Times Magazine*, 3 October 1999, pp. 38–77.

Goodman, Morris. "Serological Analysis of the Systematics of Recent Hominoids." *Human Biology* 35 (1963): 377–436.

————. "Reconstructing Human Evolution from Proteins." In *The Cambridge Encyclopedia of Human Evolution*, edited by Steve Jones et al., pp. 307–12. Cambridge: Cambridge University Press, 1992.

Goody, Jack. *The Development of the Family and Marriage in Europe*. Cambridge: Cambridge University Press, 1983.

Gordon, David C. *Self-Determination and History in the Third World*. Princeton, N.J.: Princeton University Press, 1971.

Gould, Stephen J. *Ontogeny and Phylogeny*. Cambridge, Mass.: Harvard University Press, 1977.

————. *Wonderful Life: The Burgess Shale and the Nature of History*. New York: W. W. Norton, 1989.

————. *The Structure of Evolutionary Theory*. Cambridge, Mass.: Harvard University Press, 2002.

Grady, Denise. "Exchanging Obesity's Risks for Surgery's." *New York Times*, 12 October 2000, sec. A, pp. 1, 26.

Graham, Gordon. *The Shape of the Past: A Philosophical Approach to History*. Oxford: Oxford University Press, 1997.

Gregory, Ruth W. *Anniversaries and Holidays*. 4th ed. Chicago: American Library Association, 1983.

Gribbin, John. "Human vs. Gorilla: The 1% Advantage." *Science Digest* 90 (August 1982): 73–77.

Gribbin, John, and Jeremy Cherfas. *The Monkey Puzzle: Reshaping the Evolutionary Tree.* New York: Pantheon, 1982.

Gricar, Julie M. "How Thick Is Blood? The Social Construction and Cultural Configuration of Kinship." Ph.D. diss., Columbia University, 1991.

Groves, Colin. "Human Origins." In *The First Humans: Human Origins and History to 10,000 B.C.*, edited by Göran Burenhult, pp. 33–52. New York: HarperCollins, 1993.

Groves, Colin P., and Vratislav Mazák. "An Approach to the Taxonomy of the Hominidae: Gracile Villafranchian Hominids in Africa." In *Naming Our Ancestors: An Anthology of Hominid Taxonomy*, edited by W. Eric Meikle and Sue T. Parker, pp. 107–25. Prospect Heights, Ill.: Waveland Press, 1994. Originally published in 1975.

Guare, John. *Six Degrees of Separation.* New York: Random House, 1990.

Gumbrecht, Hans U. *In 1926: Living at the Edge of Time.* Cambridge, Mass.: Harvard University Press, 1997.

Haeckel, Ernst. *Anthropogenie oder Entwickelungsgeschichte des Menschen.* Leipzig: Wilhelm Engelmann, 1874.

———. *The Evolution of Man: A Popular Exposition of the Principal Points of Human Ontogeny and Phylogeny.* New York: D. Appleton, 1879. Originally published in 1874.

Haines, Miranda, ed. *The Traveler's Handbook.* 7th ed. London: Wexas, 1997.

Halbwachs, Maurice. *The Social Frameworks of Memory.* In *Maurice Halbwachs on Collective Memory*, edited by Lewis A. Coser, pp. 37–189. Chicago: University of Chicago Press, 1992. Originally published in 1925.

———. *The Collective Memory.* New York: Harper Colophon, 1980. Originally published in 1950.

Hale, Thomas A. *Griots and Griottes: Masters of Words and Music.* Bloomington: Indiana University Press, 1998.

Haley, Alex. *Roots.* Garden City, N.Y.: Doubleday, 1976.

Hammond, Michael. "The Expulsion of the Neanderthals from Human Ancestry: Marcellin Boule and the Social Context of Scientific Research." *Social Studies of Science* 12 (1982): 1–36.

Hankiss, Agnes. "Ontologies of the Self: On the Mythological Rearranging of One's Life-History." In *Biography and Society: The Life History Approach in the Social Sciences*, edited by Daniel Bertaux, pp. 203–9. Beverly Hills, Calif.: Sage, 1981.

Hareven, Tamara K., and Kanji Masaoka. "Turning Points and Transitions: Perceptions of the Life Course." *Journal of Family History* 13 (1988): 271–89.

Hay, Robert P. "George Washington: American Moses." *American Quarterly* 21 (1969): 780–91.

Heilman, Samuel C. *The People of the Book: Drama, Fellowship, and Religion.* Chicago: University of Chicago Press, 1983.

———. *A Walker in Jerusalem.* New York: Summit Books, 1986.

Henderson, Helene, and Sue Ellen Thompson, eds. *Holidays, Festivals, and Celebrations of the World Dictionary.* 2d ed. Detroit: Omnigraphics Inc., 1997.

Henige, David P. *The Chronology of Oral Tradition: Quest for a Chimera.* London: Oxford University Press, 1974.

Herbert, Ulrich. "Good Times, Bad Times." *History Today* 36 (February 1986): 42–48.

Herman, Arthur. *The Idea of Decline in Western History.* New York: Free Press, 1997.

Herzfeld, Michael. *Ours Once More: Folklore, Ideology, and the Making of Modern Greece.* New York: Pella, 1986. Originally published in 1982.

Herzog, Hana. "The Concepts 'Old Yishuv' and 'New Yishuv' from a Sociological Perspective" (in Hebrew). *Katedra* 32 (July 1984): 99–108.

Hexter, J. H. *On Historians: Reappraisals of Some of the Makers of Modern History.* Cambridge, Mass.: Harvard University Press, 1979.

Hirst, William, and David Manier. "Remembering as Communication: A Family Recounts Its Past." In *Remembering Our Past: Studies in Autobiographical Memory,* edited by David C. Rubin, pp. 271–88. Cambridge: Cambridge University Press, 1996.

Hobsbawm, Eric J. "Introduction: Inventing Traditions." In *The Invention of Tradition,* edited by Eric J. Hobsbawm and Terence Ranger, pp. 1–14. Cambridge: Cambridge University Press, 1983.

Hoenigswald, Henry M. "Language Family Trees, Topological and Metrical." In *Biological Metaphor and Cladistic Classification: An Interdisciplinary Perspective,* edited by Henry M. Hoenigswald and Linda F. Wiener, pp. 257–67. Philadelphia: University of Pennsylvania Press, 1987.

Hoge, Warren. "Queen Breaks the Ice: Camilla's out of the Fridge." *New York Times,* 5 June 2000, sec. A, p. 4.

Holyoak, Keith J., and Paul Thagard. *Mental Leaps: Analogy in Creative Thought.* Cambridge, Mass.: MIT Press, 1995.

Hood, Andrea. "Editing the Life Course: Autobiographical Narratives, Identity Transformations, and Retrospective Framing." Unpublished manuscript, Rutgers University, Department of Sociology, 2002.

Horne, Alistair. *To Lose a Battle: France 1940.* Boston: Little, Brown & Co., 1969.

Horwitz, Allan V. *The Logic of Social Control.* New York: Plenum, 1990.

Howard, Jenna. "Memory Reconstruction in Autobiographical Narrative Construction: Analysis of the Alcoholics Anonymous Recovery Narrative." Unpublished manuscript, Rutgers University, Department of Sociology, 2000.

Howe, Stephen. *Afrocentrism: Mythical Pasts and Imagined Homes.* London: Verso, 1998.

Howells, William W. "The Dispersion of Modern Humans." In *The Cambridge Encyclopedia of Human Evolution,* edited by Steve Jones et al., pp. 389–401. Cambridge: Cambridge University Press, 1992.

Hubert, Henri. "Etude Sommaire de la Représentation du Temps dans la Religion et la Magie." In *Mélanges d'Histoire des Religions,* edited by Henri Hubert and Marcel Mauss, pp. 189–229. Paris: Félix Alcan and Guillaumin, 1909. Originally published in 1905.

Hume, David. *A Treatise of Human Nature.* London: J. M. Dent, 1977. Originally published in 1739.

Huxley, Thomas H. *Evidence as to Man's Place in Nature.* Ann Arbor: University of Michigan Press, 1959. Originally published in 1863.

Irwin-Zarecka, Iwona. *Frames of Remembrance: The Dynamics of Collective Memory.* New Brunswick, N.J.: Transaction, 1994.

Isaacson, Nicole. "The Fetus-Infant: Changing Classifications of *in-utero* Development in Medical Texts." *Sociological Forum* 11 (1996): 457–80.

James, Peter. *Centuries of Darkness: A Challenge to the Conventional Chronology of Old World Archaeology.* New Brunswick, N.J.: Rutgers University Press, 1993.

Jay, Nancy. *Throughout Your Generations Forever: Sacrifice, Religion, and Paternity.* Chicago: University of Chicago Press, 1992.

Jervis, Robert. *Perception and Misperception in International Politics.* Princeton, N.J.: Princeton University Press, 1976.

Jobs, Richard I. "The Promise of Youth: Age Categories as the Mental Framework of Rejuvenation in Postwar France." Paper presented at the Tenth Annual Interdisciplinary Conference for Graduate Scholarship, the Center for the Critical Analysis of Contemporary Culture, Rutgers University, New Brunswick, N.J., March 2000.

Johanson, Donald, and Blake Edgar. *From Lucy to Language.* New York: Simon and Schuster, 1996.

Johnson, Marshall D. *The Purpose of the Biblical Genealogies with Special Reference to the Setting of the Genealogies of Jesus.* London: Cambridge University Press, 1969.

Jones, G. I. "Time and Oral Tradition with Special Reference to Eastern Nigeria." *Journal of African History* 6 (1965): 153–60.

Jones, Gwyn. *The Norse Atlantic Saga.* 2d ed. Oxford: Oxford University Press, 1986.

Joyce, James. *Ulysses.* New York: Random House, 1986. Originally published in 1922.

Kaniel, Yehoshua. *Continuity and Change: Old Yishuv and New Yishuv during the First and Second Aliyah* (in Hebrew). Jerusalem: Yad Itzhak Ben-Zvi Publications, 1981.

Katovich, Michael A., and Carl J. Couch. "The Nature of Social Pasts and Their Use as Foundations for Situated Action." *Symbolic Interaction* 15, no. 1 (1992): 25–47.

Katriel, Tamar. *Performing the Past: A Study of Israeli Settlement Museums.* Mahwah, N.J.: Lawrence Erlbaum Associates, 1997.

Kaufman, Michael T. "Two Distinct Peoples with Two Divergent Memories Battle over One Land." *New York Times,* 4 April 1999, International section, p. 10.

Kelley, Jay. "Evolution of Apes." In *The Cambridge Encyclopedia of Human Evolution,* edited by Steve Jones et al., pp. 223–30. Cambridge: Cambridge University Press, 1992.

Kern, Stephen. *The Culture of Time and Space 1880–1918.* Cambridge, Mass.: Harvard University Press, 1983.

Khalidi, Rashid I. "What 'Final Status'?" *New York Times,* 3 October 1996, sec. A, p. 23.

Khong, Yuen F. *Analogies at War: Korea, Munich, Dien Bien Phu, and the Vietnam Decisions of 1965.* Princeton, N.J.: Princeton University Press, 1992.

Kifner, John. "Israeli and Palestinian Leaders Vow to Keep Working for Peace." *New York Times,* 27 July 2000, sec. A, pp. 1, 11.

Kintsch, Walter, and Edith Greene. "The Role of Culture-Specific Schemata in the Comprehension and Recall of Stories." *Discourse Processes* 1 (1978): 1–13.

Klaatsch, Hermann. *The Evolution and Progress of Mankind.* New York: Frederick A. Stokes, 1923.

Koerner, Konrad. "On Schleicher and Trees." In *Biological Metaphor and Cladistic Classifi-*

cation: An Interdisciplinary Perspective, edited by Henry M. Hoenigswald and Linda F. Wiener, pp. 109–13. Philadelphia: University of Pennsylvania Press, 1987.

Koonz, Claudia. "Between Memory and Oblivion: Concentration Camps in German Memory." In *Commemorations: The Politics of National Identity,* edited by John R. Gillis, pp. 258–80. Princeton, N.J.: Princeton University Press, 1994.

Koselleck, Reinhart. "Modernity and the Planes of Historicity." In *Futures Past: On the Semantics of Historical Time,* pp. 3–20. Cambridge, Mass.: MIT Press, 1985. Originally published in 1968.

Krauss, Clifford. "Son of the Poor Is Elected in Peru over Ex-President." *New York Times,* 4 June 2001, sec. A, pp. 1, 6.

Kristof, Nicholas D. "With Genghis Revived, What Will Mongols Do?" *New York Times,* 23 March 1990, International section, p. A4.

Kubler, George. *The Shape of Time.* New Haven, Conn.: Yale University Press, 1962.

Lamarck, Jean-Baptiste. *Zoological Philosophy: An Exposition with Regard to the Natural History of Animals.* New York: Hafner, 1963. Originally published in 1809.

Landau, Misia. "Human Evolution as Narrative." *American Scientist* 72 (1984): 262–68.

Lasch, Christopher. *The True and Only Heaven: Progress and Its Critics.* New York: W. W. Norton, 1991.

Leach, Edmund. "Two Essays concerning the Symbolic Representation of Time." In *Rethinking Anthropology,* pp. 124–36. London: Athlone, 1961.

———. "On Certain Unconsidered Aspects of Double Descent Systems." *Man* 62 (1962): 130–34.

Leakey, Louis S. B., P. V. Tobias, and J. R. Napier. "A New Species of the Genus *Homo* from Olduvai Gorge." In *Naming Our Ancestors: An Anthology of Hominid Taxonomy,* edited by W. Eric Meikle and Sue T. Parker, pp. 94–101. Prospect Heights, Ill.: Waveland Press, 1994. Originally published in 1964.

Lévi-Strauss, Claude. *The Savage Mind.* Chicago: University of Chicago Press, 1966. Originally published in 1962.

Lewis, Bernard. *History: Remembered, Recovered, Invented.* Princeton, N.J.: Princeton University Press, 1975.

Lewis, Martin W., and Kären E. Wigen. *The Myth of Continents: A Critique of Metageography.* Berkeley and Los Angeles: University of California Press, 1997.

Lewontin, Richard. *Human Diversity.* New York: Scientific American Books, 1982.

Libove, Jessica. "Guardians of Collective Memory: The Mnemonic Functions of the Griot in West Africa." Unpublished manuscript, Rutgers University, Department of Anthropology, 2000.

Lipson, Marjorie Y. "The Influence of Religious Affiliation on Children's Memory for Text Information." *Reading Research Quarterly* 18 (1983): 448–57.

Loewen, James W. *Lies My Teacher Told Me: Everything Your American History Textbook Got Wrong.* New York: Touchstone, 1996.

Lorenz, Konrad. *On Aggression.* New York: Bantam, 1971. Originally published in 1963.

Lovejoy, Arthur O. *The Great Chain of Being: A Study of the History of an Idea.* Cambridge, Mass.: Harvard University Press, 1936.

———. "The Argument for Organic Evolution before the Origin of Species, 1830–1858."

In *Forerunners of Darwin: 1745–1859*, edited by Bentley Glass et al., pp. 356–414. Baltimore: Johns Hopkins University Press, 1959.

Lowenstein, Jerold, and Adrienne Zihlman. "The Invisible Ape." *New Scientist*, 3 December 1988, pp. 56–59.

Lowenthal, David. *The Past Is a Foreign Country*. Cambridge: Cambridge University Press, 1985.

————. *Possessed by the Past: The Heritage Crusade and the Spoils of History*. New York: Free Press, 1996.

Lynch, Kevin. *What Time Is This Place?* Cambridge, Mass.: MIT Press, 1972.

Maclear, Michael. *The Ten Thousand Day War: Vietnam, 1945–1975*. Toronto: Methuen, 1981.

Malcolm, Noel. *Kosovo: A Short History*. Updated ed. New York: HarperPerennial, 1999.

Malkki, Lisa H. *Purity and Exile: Violence, Memory, and National Cosmology among Hutu Refugees in Tanzania*. Chicago: University of Chicago Press, 1995.

Mandler, Jean M. *Stories, Scripts, and Scenes: Aspects of Schema Theory*. Hillsdale, N.J.: Lawrence Erlbaum, 1984.

Mannheim, Karl. "The Problem of Generations." In *Essays on the Sociology of Knowledge*, pp. 276–320. London: Routledge and Kegan Paul, 1951. Originally published in 1927.

————. *Ideology and Utopia: An Introduction to the Sociology of Knowledge*. New York: Harvest Books, 1936. Originally published in 1929.

Martin, Robert. "Classification and Evolutionary Relationships." In *The Cambridge Encyclopedia of Human Evolution*, edited by Steve Jones et al., pp. 17–23. Cambridge: Cambridge University Press, 1992.

Marx, Karl. "The Eighteenth Brumaire of Louis Bonaparte." In *The Marx-Engels Reader*, edited by Robert C. Tucker, 2d ed., pp. 594–617. New York: W. W. Norton, 1978. Originally published in 1852.

May, Ernest R. *"Lessons" of the Past: The Use and Misuse of History in American Foreign Policy*. New York: Oxford University Press, 1973.

Mayr, Ernst. "Taxonomic Categories in Fossil Hominids." In *Naming Our Ancestors: An Anthology of Hominid Taxonomy*, edited by Eric Meikle and Sue T. Parker, pp. 152–70. Prospect Heights, Ill.: Waveland Press, 1994. Originally published in 1950.

————. "The Taxonomic Evaluation of Fossil Hominids." In *Climbing Man's Family Tree: A Collection of Major Writings on Human Phylogeny, 1699 to 1971*, edited by Theodore D. McCown and Kenneth A. R. Kennedy, pp. 372–86. Englewood Cliffs, N.J.: Prentice-Hall, 1972. Originally published in 1963.

McAdams, Dan P. *The Stories We Live By: Personal Myths and the Making of the Self*. New York: William Morrow, 1993.

McCain, John. Interview by Bob Edwards. *Morning News*. National Public Radio, 14 September 2001.

McCown, Theodore D., and Kenneth A. R. Kennedy, eds. *Climbing Man's Family Tree: A Collection of Major Writings on Human Phylogeny, 1699 to 1971*. Englewood Cliffs, N.J.: Prentice-Hall, 1972.

McNeil, Kenneth, and James D. Thompson. "The Regeneration of Social Organizations." *American Sociological Review* 36 (1971): 624–37.

Meikle, W. Eric, and Sue T. Parker. "Introduction: Names, Binomina, and Nomenclature in

Paleoanthropology." In *Naming Our Ancestors: An Anthology of Hominid Taxonomy*, pp. 1–18. Prospect Heights, Ill.: Waveland Press, 1994.

Milgram, Stanley. "The Small World Problem." In *The Individual in a Social World: Essays and Experiments*, 2d ed., pp. 259–75. New York: McGraw-Hill, 1992. Originally published in 1967.

Miller, Joseph C. "Introduction: Listening for the African Past." In *The African Past Speaks: Essays on Oral Tradition and History*, pp. 1–59. Folkestone, England: William Dawson, 1980.

Milligan, Melinda J. "Interactional Past and Potential: The Social Construction of Place Attachment." *Symbolic Interaction* 21 (1998): 1–33.

"Montenegro Asks Forgiveness from Croatia." *New York Times*, 25 June 2000, International section, p. 9.

Morgan, Lewis H. *Systems of Consanguinity and Affinity of the Human Family*. Lincoln: University of Nebraska Press, 1997. Originally published in 1871.

Morris, Desmond. *The Naked Ape*. New York: McGraw-Hill, 1967.

Morris, Ian. "Periodization and the Heroes: Inventing a Dark Age." In *Inventing Ancient Culture: Historicism, Periodization, and the Ancient World*, edited by Mark Golden and Peter Toohey, pp. 96–131. London: Routledge, 1997.

Morris, Ramona, and Desmond Morris. *Men and Apes*. New York: Bantam, 1968. Originally published in 1966.

Mullaney, Jamie. "Making It 'Count': Mental Weighing and Identity Attribution." *Symbolic Interaction* 22 (1999): 269–83.

————. "Like A Virgin: Temptation, Resistance, and the Construction of Identities Based on 'Not Doings.'" *Qualitative Sociology* 24 (2001): 3–24.

Murchie, Guy. *The Seven Mysteries of Life: An Exploration in Science and Philosophy*. New York: Mariner Books, 1999. Originally published in 1978.

Mydans, Seth. "Cambodian Leader Resists Punishing Top Khmer Rouge." *New York Times*, 29 December 1998, sec. A, p. 1.

————. "Under Prodding, 2 Apologize for Cambodian Anguish." *New York Times*, 30 December 1998, International section.

Neustadt, Richard E., and Ernest R. May. *Thinking in Time: The Uses of History for Decision-Makers*. New York: Free Press, 1986.

Nora, Pierre. "Between Memory and History: Les Lieux de Memoire." *Representations* 26 (1989): 7–25.

Nuttall, George H. *Blood Immunity and Blood Relationship: A Demonstration of Certain Blood-Relationships amongst Animals by means of the Precipitin Test for Blood*. Cambridge: Cambridge University Press, 1904.

Ojito, Mirta. "Blacks on a Brooklyn Street: Both Cynics and Optimists Speak Out." *New York Times*, 26 March 1998, International section, p. A13.

Onishi, Norimitsu. "A Tale of the Mullah and Muhammad's Amazing Cloak." *New York Times*, 19 December 2001, sec. B, pp. 1–3.

Oppenheimer, Jane M. "Haeckel's Variations on Darwin." In *Biological Metaphor and Cladistic Classification: An Interdisciplinary Perspective*, edited by Henry M. Hoenigswald and Linda F. Wiener, pp. 123–35. Philadelphia: University of Pennsylvania Press, 1987.

Packard, Vance. *The Waste Makers.* New York: David McKay, 1960.

Park, Robert E., and Ernest W. Burgess. *Introduction to the Science of Sociology.* Abridged ed. Chicago: University of Chicago Press, 1969. Originally published in 1921.

Parsons, Talcott. "The Kinship System of the Contemporary United States." In *Essays in Sociological Theory,* pp. 177–96. Rev. ed. New York: Free Press, 1964. Originally published in 1943.

Peirce, Charles S. *Collected Papers of Charles Sanders Peirce.* Cambridge, Mass.: Harvard University Press, 1962. Originally published in 1932.

Perring, Stefania, and Dominic Perring. *Then and Now.* New York: Macmillan, 1991.

Pillemer, David B. *Momentous Events, Vivid Memories.* Cambridge, Mass.: Harvard University Press, 1998.

Polacco, Patricia. *Pink and Say.* New York: Philomel Books, 1994.

Pool, Ithiel de Sola, and Manfred Kochen. "Contacts and Influence." In *The Small World,* edited by Manfred Kochen, pp. 3–51. Norwood, N.J.: Ablex, 1989. Originally published in 1978.

Popkin, Richard H. "The Pre-Adamite Theory in the Renaissance." In *Philosophy and Humanism: Renaissance Essays in Honor of Paul Oskar Kristeller,* edited by Edward P. Mahoney, pp. 50–69. New York: Columbia University Press, 1976.

Powell, Bill. "The Innocents of WW II?" *Newsweek,* 12 December 1994, pp. 52–53.

Pritchard, Robert. "The Effects of Cultural Schemata on Reading Processing Strategies." *Reading Research Quarterly* 25 (1990): 273–95.

Purcell, Kristen. "Leveling the Playing Field: Constructing Parity in the Modern World." Ph.D. diss., Rutgers University, 2001.

Radcliffe-Brown, Alfred R. "Patrilineal and Matrilineal Succession." In *Structure and Function in Primitive Society,* pp. 32–48. New York: Free Press, 1965. Originally published in 1935.

————. "The Study of Kinship Systems." In *Structure and Function in Primitive Society,* pp. 49–89. New York: Free Press, 1965. Originally published in 1941.

Reader, John. *Missing Links: The Hunt for Earliest Man.* Boston: Little, Brown, & Co., 1981.

Remnick, David. "The Outsider." *The New Yorker,* 25 May 1998, pp. 80–95.

Rendsburg, Gary A. "Archeological Fairness." *New York Times,* 11 October 1996, sec. A, p. 38.

Renfrew, Colin. *Archaeology and Language: The Puzzle of Indo-European Origins.* New York: Cambridge University Press, 1987.

Richards, Robert J. *The Meaning of Evolution: The Morphological Construction and Ideological Reconstruction of Darwin's Theory.* Chicago: University of Chicago Press, 1992.

Ritvo, Harriet. "Border Trouble: Shifting the Line between People and Other Animals." *Social Research* 62 (1995): 481–500.

Robinson, John A. "First Experience Memories: Contexts and Functions in Personal Histories." In *Theoretical Perspectives on Autobiographical Memory,* edited by Martin A. Conway et al., pp. 223–39. Dordrecht: Kluwer Academic Publishers, 1992.

Rogers, Alan R., and Lynn B. Jorde. "Genetic Evidence on Modern Human Origins." *Human Biology* 67 (1995): 1–36.

Rolston, Bill. *Drawing Support: Murals in the North of Ireland.* Belfast: Beyond the Pale Publications, 1992.

Rosenberg, Harold. *Saul Steinberg.* New York: Alfred A. Knopf, 1978.

Rosenblum, Doron. "Because Somebody Needs to Be an Israeli in Israel" (in Hebrew). *Ha'aretz,* 29 April 1998, Independence Day Supplement.

Rubinstein, Amnon. *To Be a Free People* (in Hebrew). Tel-Aviv: Schocken, 1977.

Ruvolo, Maryellen, et al. "Mitochondrial COII Sequences and Modern Human Origins." *Molecular Biology and Evolution* 10 (1993): 1115–35.

Sachs, Susan. "Bin Laden Images Mesmerize Muslims." *New York Times,* 9 October 2001, sec. B, p. 6.

Sahlins, Marshall. *Historical Metaphors and Mythical Realities: Structure in the Early History of the Sandwich Islands Kingdom.* Ann Arbor: University of Michigan Press, 1981.

Sarich, Vincent. "Immunological Evidence on Primates." In *The Cambridge Encyclopedia of Human Evolution,* edited by Steve Jones et al., pp. 303–6. Cambridge: Cambridge University Press, 1992.

Sarich, Vincent, and Allan C. Wilson. "Immunological Time Scale for Hominid Evolution." *Science* 158 (1967): 1200–1203.

Saussure, Ferdinand de. *Course in General Linguistics.* New York: Philosophical Library, 1959. Originally published in 1915.

Schank, Roger C., and Robert P. Abelson. "Scripts, Plans, and Knowledge." In *Thinking: Readings in Cognitive Science,* edited by P. N. Johnson-Laird and P. C. Wason, pp. 421–32. Cambridge: Cambridge University Press, 1977.

Schiff, Zeev, and Eitan Haber, eds. *Israel, Army, and Defense: A Dictionary* (in Hebrew). Jerusalem: Zmora, Bitan, Modan, 1976.

Schmitt, Raymond L. "Symbolic Immortality in Ordinary Contexts: Impediments to the Nuclear Era." *Omega* 13 (1982–83): 95–116.

Schneider, David M. *American Kinship: A Cultural Account.* Englewood Cliffs, N.J.: Prentice-Hall, 1968.

Schneider, Herbert W., and Shepard B. Clough. *Making Fascists.* Chicago: University of Chicago Press, 1929.

Schuman, Howard, and Cheryl Rieger. "Historical Analogies, Generational Effects, and Attitudes toward War." *American Sociological Review* 57 (1992): 315–26.

Schuman, Howard, and Jacqueline Scott. "Generations and Collective Memories." *American Sociological Review* 54 (1989): 359–81.

Schusky, Ernest L. *Variation in Kinship.* New York: Holt, Rinehart, and Winston, 1974.

Schutz, Alfred. "Phenomenology and the Social Sciences." In *Collected Papers, vol. 1: The Problem of Social Reality,* edited by Maurice Natanson, pp. 118–39. The Hague: Martinus Nijhoff, 1973. Originally published in 1940.

———. "Making Music Together: A Study in Social Relationship." In *Collected Papers, vol. 2: Studies in Social Theory,* edited by Arvid Brodersen, pp. 159–78. The Hague: Martinus Nijhoff, 1964. Originally published in 1951.

Schutz, Alfred, and Thomas Luckmann. *The Structures of the Life-World.* Evanston, Ill.: Northwestern University Press, 1973.

Schwartz, Barry. *Vertical Classification: A Study in Structuralism and the Sociology of Knowl-edge*. Chicago: University of Chicago Press, 1981.

————. "The Social Context of Commemoration: A Study in Collective Memory." *Social Forces* 61 (1982): 374–96.

Scott 1999 Standard Postage Stamp Catalogue. Sidney, Ohio: Scott Publishing Co., 1998.

Secord, James A. Introduction to *Vestiges of the Natural History of Creation*, by Robert Cham-bers, pp. ix—xlv. Chicago: University of Chicago Press, 1994.

Shils, Edward. *Tradition*. Chicago: University of Chicago Press, 1981.

Shoumatoff, Alex. *The Mountain of Names: A History of the Human Family*. New York: Simon and Schuster, 1985.

Sibley, Charles G. "DNA-DNA Hybridisation in the Study of Primate Evolution." In *The Cambridge Encyclopedia of Human Evolution*, edited by Steve Jones et al., pp. 313–15. Cambridge: Cambridge University Press, 1992.

Silberman, Neil A. *Between Past and Present: Archaeology, Ideology, and Nationalism in the Modern Middle East*. New York: Henry Holt, 1989.

Silver, Ira. "Role Transitions, Objects, and Identity." *Symbolic Interaction* 19, no. 1 (1996): 1–20.

Simmel, Georg. "The Persistence of Social Groups." *American Journal of Sociology* 3 (1897–98): 662–98.

————. *The Sociology of Georg Simmel*, edited by Kurt H. Wolff. New York: Free Press, 1950. Originally published in 1908.

————. "Written Communication." In *The Sociology of Georg Simmel*, edited by Kurt H. Wolff, pp. 352–55. New York: Free Press, 1950. Originally published in 1908.

————. "Bridge and Door." *Theory, Culture & Society* 11 (1994): 5–10. Originally published in 1909.

Simpson, George G. *Principles of Animal Taxonomy*. New York: Columbia University Press, 1961.

————. "The Meaning of Taxonomic Statements." In *Naming Our Ancestors: An Anthology of Hominid Taxonomy*, edited by W. Eric Meikle and Sue T. Parker, pp. 172–206. Prospect Heights, Ill.: Waveland, 1994. Originally published in 1963.

Simpson, Ruth. "I Was There: Establishing Ownership of Historical Moments." Paper pre-sented at the Annual Meeting of the American Sociological Association, Los Angeles, 1994.

————. "Microscopic Worlds, Miasmatic Theories, and Myopic Vision: Changing Con-ceptions of Air and Social Space." Paper presented at the Annual Meeting of the Amer-ican Sociological Association, Chicago, 1999.

Smith, Anthony D. *The Ethnic Origins of Nations*. Oxford: Basil Blackwell, 1986.

Smith, Jason S. "The Strange History of the Decade: Modernity, Nostalgia, and the Perils of Periodization." *Journal of Social History* 32 (1998): 263–85.

Snodgrass, A. M. *The Dark Age of Greece: An Archeological Survey of the Eleventh to the Eighth Centuries B.C.* Edinburgh: Edinburgh University Press, 1971.

Sokolowski, S. Wojciech. "Historical Tradition in the Service of Ideology." *Conjecture* (September 1992): 4–11.

Sorabji, Richard. *Time, Creation, and the Continuum: Theories in Antiquity and the Early Middle Ages.* Ithaca, N.Y.: Cornell University Press, 1983.

Sorokin, Pitirim A. *Sociocultural Causality, Space, Time: A Study of Referential Principles of Sociology and Social Science.* Durham, N.C.: Duke University Press, 1943.

Sorokin, Pitirim A., and Robert K. Merton. "Social Time: A Methodological and Functional Analysis." *American Journal of Sociology* 42 (1937): 615–29.

"Special Purim." *Encyclopaedia Judaica* 13: 1396–1400. Jerusalem: Keter, 1972.

Spencer, Herbert. *Principles of Sociology.* Hamden, Conn.: Archon, 1969. Originally published in 1876.

Spillman, Lyn. *Nation and Commemoration: Creating National Identities in the United States and Australia.* Cambridge: Cambridge University Press, 1997.

—————. "When Do Collective Memories Last? Founding Moments in the United States and Australia." *Social Science History* 22 (1998): 445–77.

Steffensen, Margaret S., Chitra Joag-Dev, and Richard C. Anderson. "A Cross-Cultural Perspective on Reading Comprehension." *Reading Research Quarterly* 15 (1979): 10–29.

Stepan, Nancy. *The Idea of Race in Science: Great Britain 1800–1960.* Hamden, Conn.: Archon Books, 1982.

Stocking, George W. "French Anthropology in 1800." In *Race, Culture, and Evolution: Essays in the History of Anthropology,* pp. 15–41. New York: Free Press, 1968. Originally published in 1964.

—————. "The Dark-Skinned Savage: The Image of Primitive Man in Evolutionary Anthropology." In *Race, Culture, and Evolution: Essays in the History of Anthropology,* pp. 112–32. New York: Free Press, 1968.

—————. "The Persistence of Polygenist Thought in Post-Darwinian Anthropology." In *Race, Culture, and Evolution: Essays in the History of Anthropology,* pp. 44–68. New York: Free Press, 1968.

Stovel, Katherine. "The Malleability of Precedent." Paper presented at the Annual Meeting of the Social Science History Association, New Orleans, 1996.

Strauss, Anselm L. *Mirrors and Masks: The Search for Identity.* London: Martin Robertson, 1977.

Strauss, Barry S. "The Problem of Periodization: The Case of the Peloponnesian War." In *Inventing Ancient Culture: Historicism, Periodization, and the Ancient World,* edited by Mark Golden and Peter Toohey, pp. 165–75. London: Routledge, 1997.

Stringer, Christopher. "Evolution of Early Humans." In *The Cambridge Encyclopedia of Human Evolution,* edited by Steve Jones et al., pp. 241–51. Cambridge: Cambridge University Press, 1992.

Stringer, Christopher, and Robin McKie. *African Exodus: The Origins of Modern Humanity.* New York: Henry Holt, 1997. Originally published in 1996.

Swadesh, Morris. "What Is Glottochronology?" In *The Origin and Diversification of Language,* pp. 271–84. Chicago: Aldine, 1971. Originally published in 1960.

Tattersall, Ian. "Species Recognition in Human Paleontology." In *Naming Our Ancestors: An Anthology of Hominid Taxonomy,* edited by W. Eric Meikle and Sue T. Parker, pp. 240–54. Prospect Heights, Ill.: Waveland Press, 1994. Originally published in 1986.

————. *The Fossil Trail: How We Know What We Think We Know about Human Evolution.* New York: Oxford University Press, 1995.

Tattersall, Ian, and Jeffrey H. Schwartz. *Extinct Humans.* Boulder, Colo.: Westview, 2000.

Temkin, Owsei. "The Idea of Descent in Post-Romantic German Biology: 1848–1858." In *Forerunners of Darwin: 1745–1859,* edited by Bentley Glass et al., pp. 323–55. Baltimore: Johns Hopkins University Press, 1959.

Thomas, Evan. "The Road to September 11." *Newsweek,* 1 October 2001, pp. 38–49.

Thomas, Northcote W. *Kinship Organisations and Group Marriage in Australia.* New York: Humanities Press, 1966. Originally published in 1906.

Thorne, Alan G., and Milford H. Wolpoff. "Regional Continuity in Australasian Pleistocene Hominid Evolution." *American Journal of Physical Anthropology* 55 (1981): 337–49.

Toffler, Alvin. *Future Shock.* New York: Random House, 1970.

Tönnies, Ferdinand. *Community and Society.* New York: Harper Torchbooks, 1963. Originally published in 1887.

Topinard, Paul. *Anthropology.* London: Chapman & Hall, 1878.

Traas, Wendy. "Turning Points and Defining Moments: An Exploration of the Narrative Styles That Structure the Personal and Group Identities of Born-Again Christians and Gays and Lesbians." Unpublished manuscript, Rutgers University, Department of Sociology, 2000.

Trevor-Roper, Hugh. "The Invention of Tradition: The Highland Tradition of Scotland." In *The Invention of Tradition,* edited by Eric J. Hobsbawm and Terence Ranger, pp. 15–41. Cambridge: Cambridge University Press, 1983.

Turner, Victor. "Betwixt and Between: The Liminal Period in Rites de Passage." In *The Forest of Symbols: Aspects of Ndembu Ritual,* pp. 93–111. Ithaca, N.Y.: Cornell University Press, 1970. Originally published in 1964.

Twine, France W. *Racism in a Racial Democracy: The Maintenance of White Supremacy in Brazil.* New Brunswick, N.J.: Rutgers University Press, 1998.

Van Gennep, Arnold. *The Rites of Passage.* Chicago: University of Chicago Press, 1960. Originally published in 1908.

Van Natta, Don, and James Risen. "Investigators Say Evidence Points to Bin Laden Aides as Planning Attack." *New York Times,* 8 October 2001, sec. B, p. 7.

Vansina, Jan. *Oral Tradition as History.* Madison: University of Wisconsin Press, 1985.

Verdery, Katherine. *The Political Lives of Dead Bodies: Reburial and Postsocialist Change.* New York: Columbia University Press, 1999.

Vinitzky-Seroussi, Vered. *After Pomp and Circumstance: High School Reunion as an Autobiographical Occasion.* Chicago: University of Chicago Press, 1998.

————. "Commemorating a Difficult Past: Yitzhak Rabin's Memorials." *American Sociological Review* 67 (2002): 30–51.

Vogt, Carl. *Lectures on Man: His Place in Creation and in the History of the Earth.* London: Longman, Green, Longman, and Roberts, 1864.

Wachter, Kenneth W. "Ancestors at the Norman Conquest." In *Genealogical Demography,* edited by Bennett Dyke and Warren T. Morrill, pp. 85–93. New York: Academic Press, 1980.

Wade, Nicholas. "To People the World, Start With 500." *New York Times*, 11 November 1997, sec. F, pp. 1–3.

————. "The Human Family Tree: 10 Adams and 18 Eves." *New York Times*, 2 May 2000, sec. F, pp. 1–5.

————. "The Origin of the Europeans." *New York Times*, 14 November 2000, sec. F, pp. 1–9.

Wagner, Anthony. "Bridges to Antiquity." In *Pedigree and Progress: Essays in the Genealogical Interpretation of History*, pp. 50–75. London: Phillimore, 1975.

Wagner-Pacifici, Robin. *Theorizing the Standoff: Contingency in Action*. Cambridge: Cambridge University Press, 2000.

Walzer, Michael. *Exodus and Revolution*. New York: Basic Books, 1984.

Warner, W. Lloyd. *The Living and the Dead*. New Haven, Conn.: Yale University Press, 1959.

————. *The Family of God*. New Haven, Conn.: Yale University Press, 1961.

Waters, Mary C. *Ethnic Options: Choosing Identities in America*. Berkeley and Los Angeles: University of California Press, 1990.

Weaver, Robert S. *International Holidays: 204 Countries from 1994 through 2015*. Jefferson, N.C.: McFarland, 1995.

Weber, Max. *Economy and Society: An Outline of Interpretive Sociology*. Berkeley and Los Angeles: University of California Press, 1978. Originally published in 1925.

Weidenreich, Franz. "Facts and Speculations concerning the Origin of *Homo sapiens*." In *Climbing Man's Family Tree: A Collection of Major Writings on Human Phylogeny, 1699 to 1971*, edited by Theodore D. McCown and Kenneth A. R. Kennedy, pp. 336–53. Englewood Cliffs, N.J.: Prentice-Hall, 1972. Originally published in 1947.

Wells, Rulon S. "The Life and Growth of Language: Metaphors in Biology and Linguistics." In *Biological Metaphor and Cladistic Classification: An Interdisciplinary Perspective*, edited by Henry M. Hoenigswald and Linda F. Wiener, pp. 39–80. Philadelphia: University of Pennsylvania Press, 1987.

Werner, Heinz. *Comparative Psychology of Mental Development*. Rev. ed. New York: International Universities Press, 1957.

West, Elliott. "A Longer, Grimmer, but More Interesting Story." In *Trails toward a New Western History*, edited by Patricia Nelson Limerick et al., pp. 103–11. Lawrence: University Press of Kansas, 1991.

White, Hayden. "The Historical Text as Literary Artifact." In *Tropics of Discourse: Essays in Cultural Criticism*, pp. 81–99. Baltimore: Johns Hopkins University Press, 1978. Originally published in 1974.

Wilcox, Donald J. *The Measure of Times Past: Pre-Newtonian Chronologies and the Rhetoric of Relative Time*. Chicago: University of Chicago Press, 1987.

Wilford, John N. "When Humans Became Human." *New York Times*, 26 February 2002, sec. F, pp. 1, 5.

————. "A Fossil Unearthed in Africa Pushes Back Human Origins." *New York Times*, 11 July 2002, sec. A, pp. 1, 12.

Wilson, Ian. *The Shroud of Turin: The Burial Cloth of Jesus Christ?* Garden City, N.Y.: Doubleday, 1978.

Winnicott, D. W. "Transitional Objects and Transitional Phenomena." In *Playing and Reality*, pp. 1–25. London: Tavistock, 1971. Originally published in 1953.

Wise, Michael Z. "Idea of a Unified Cultural Heritage Divides Europe." *New York Times*, 29 January 2000, sec. B, pp. 9–11.

Wolpoff, Milford H., et al. "Modern Human Origins." *Science* 241 (1988): 772–73.

Wood, Bernard A. "Evolution and Australopithecines." In *The Cambridge Encyclopedia of Human Evolution*, edited by Steve Jones et al., pp. 231–40. Cambridge: Cambridge University Press, 1992.

Woodward, Kenneth L. "2000 Years of Jesus." *Newsweek*, 29 March 1999, pp. 52–55.

Yerushalmi, Yosef H. *Zakhor: Jewish History and Jewish Memory*. Seattle: University of Washington Press, 1982.

Young, Marilyn B. *The Vietnam Wars 1945–1990*. New York: HarperCollins, 1991.

Zerubavel, Eviatar. "The French Republican Calendar: A Case Study in the Sociology of Time." *American Sociological Review* 42 (1977): 868–77.

———. *Patterns of Time in Hospital Life: A Sociological Perspective*. Chicago: University of Chicago Press, 1979.

———. "If Simmel Were a Fieldworker: On Formal Sociological Theory and Analytical Field Research." *Symbolic Interaction* 3, no. 2 (1980): 25–33.

———. *Hidden Rhythms: Schedules and Calendars in Social Life*. Chicago: University of Chicago Press, 1981.

———. "Easter and Passover: On Calendars and Group Identity." *American Sociological Review* 47 (1982): 284–89.

———. "Personal Information and Social Life." *Symbolic Interaction* 5, no. 1 (1982): 97–109.

———. *The Seven-Day Circle: The History and Meaning of the Week*. New York: Free Press, 1985.

———. *The Fine Line: Making Distinctions in Everyday Life*. New York: Free Press, 1991.

———. *Terra Cognita: The Mental Discovery of America*. New Brunswick, N.J.: Rutgers University Press, 1992.

———. "In the Beginning: Notes on the Social Construction of Historical Discontinuity." *Sociological Inquiry* 63 (1993): 457–59.

———. "Lumping and Splitting: Notes on Social Classification." *Sociological Forum* 11 (1996): 421–33.

———. *Social Mindscapes: An Invitation to Cognitive Sociology*. Cambridge, Mass.: Harvard University Press, 1997.

———. "Language and Memory: 'Pre-Columbian' America and the Social Logic of Periodization." *Social Research* 65 (1998): 315–30.

———. *The Clockwork Muse: A Practical Guide to Writing Theses, Dissertations, and Books*. Cambridge, Mass.: Harvard University Press, 1999.

———. "The Elephant in the Room: Notes on the Social Organization of Denial." In *Culture in Mind: Toward a Sociology of Culture and Cognition*, edited by Karen A. Cerulo, pp. 21–27. New York: Routledge, 2002.

———. "Calendars and History: A Comparative Study of the Social Organization of National Memory." In *States of Memory: Conflicts, Continuities, and Transformations in Na-*

tional Commemoration, edited by Jeffrey K. Olick. Durham, N.C.: Duke University Press, in press.

—————. "The Social Marking of the Past: Toward a Socio-Semiotics of Memory." In *The Cultural Turn*, edited by Roger Friedland and John Mohr. Cambridge: Cambridge University Press, in press.

Zerubavel, Yael. "The Death of Memory and the Memory of Death: Masada and the Holocaust as Historical Metaphors." *Representations* 45 (winter 1994): 72–100.

—————. *Recovered Roots: Collective Memory and the Making of Israeli National Tradition*. Chicago: University of Chicago Press, 1995.

—————. "The Forest as a National Icon: Literature, Politics, and the Archeology of Memory." *Israel Studies* 1 (1996): 60–99.

—————. "Travels in Time and Space: Legendary Literature as a Vehicle for Shaping Collective Memory" (in Hebrew). *Teorya Uviqoret* 10 (summer 1997): 69–80.

—————. "The Mythological Sabra and the Jewish Past: Trauma, Memory, and Contested Identities." *Israel Studies* 7 (2002).

—————. *Desert Images: Visions of the Counter-Place in Israeli Culture*. Chicago: University of Chicago Press, forthcoming.

Zielbauer, Paul. "Found in Clutter, a Relic of Lincoln's Death." *New York Times*, 5 July 2001, sec. A, p. 1—sec. B, p. 5.

Zussman, Robert. "Autobiographical Occasions: Photography and the Representation of the Self." Paper presented at the Annual Meeting of the American Sociological Association, Chicago, August 1999.

Author Index

Subject Index